Genetic Engineering - Dream or Nightmare?

The Brave New World of Bad Science and Big Business

MAE-WAN HO

Gateway Books, Bath, UK

First published in 1998
by GATEWAY BOOKS
The Hollies, Wellow,
Bath BA2 8QJ, UK

Distributed in the USA by
ACCESS PUBLISHERS NETWORK
6893 Sullivan Rd, Grawn, MI 49637

Set in 10½ on 13 Bookman,
by Synergie, Bristol

Printed and bound by
Redwood of Trowbridge

Cover design by Synergie, Bristol

British Library Cataloguing in Publication Data
A catalogue record for this book is available from the British Library

Hardcover edition ISBN 1 85860 052 9
Paperback edition ISBN 1 85860 051 0

Foreword

In July, 1994, I attended a meeting that changed my life. It was a conference in Penang, Malaysia, with the enigmatic title, *Redefining the Life Sciences*, and had been organised by Vandana Shiva, Martin Khor, Tewolde Egziagher and my colleague at the Open University, Brian Goodwin.

I was invited, along with fifty other participants, consisting of scientists, social scientists, policy-makers and political activists, drawn from fifteen countries. I learned how the dominant reductionist scientific world view of the West was destroying the earth, creating poverty and suffering for vast numbers of people. At the same time, I learned how those same people were regenerating the land and their livelihoods with indigenous wisdom accumulated and passed down through the ages. Within a day, we felt like friends who had lost touch for a hundred years. We drafted a *Scientists' Statement on the Need for Greater Regulation and Control of Genetic Engineering* which was published by the Third World Network and widely distributed.

As a result of that inspiring meeting, I began to realise how much science matters in the affairs of the real world, not just in terms of practical inventions like genetic engineering, but in how that scientific world view takes hold of people's unconscious, making them act, unthinkingly, to shape the world to the detriment of human beings, how that science is used, often without conscious intent, to intimidate and control, how it is used to obfuscate, to exploit and oppress. I began to see how that dominant world view generates a selective blindness in scientists themselves, making them ignore scientific evidence, or fail to interpret them correctly. I have felt obliged, ever since, to tell the other side of the story, in the interest of promoting real public understanding of science in general, and of genetic engineering biotechnology in particular. Earlier this year, Martin Khor finally convinced me to write this book, by skilfully sketching out its structure for me.

This book argues the case against genetic engineering biotechnology as bad science working together with big business for quick profit, against the public good, against public will and aspirations, against the moral values of society and the world community. I show how bad science, in the form of genetic determinism, is at odds with the reality of scientific findings. How it gives rise to misguided practices and projects in genetic engineering biotechnology that are unethical and exploitative. I show how the very failures of genetic engineering biotechnology stem from the inability (of reductionist thinking) to take account of complexity, interconnectedness and wholeness. Most of all, it reveals how that same genetic determinist mindset leads scientists to ignore or misread existing scientific evidence, already strongly suggesting that genetic engineering biotechnology is *inherently* hazardous to human and animal health and the ecological environment.

It is a matter of urgency that an immediate moratorium should be imposed on further releases and marketing of genetically engineered products, pending an independent public enquiry into the hazards and risks of genetic engineering biotechnology which takes into proper account the most comprehensive body of scientific findings, as well as the social, moral implications.

This book owes its existence to so many of my friends that I hope they will forgive me if I fail to mention all of their names. Edward Goldsmith and Peter Bunyard initiated me to Gaia and global ecology and gave me the benefit of their extensive knowledge. Martin Khor, Vandana Shiva, Chee Yoke Ling, Tewolde Egziagher and Gurdial Nijar of the Third World Network all taught me a great deal about world politics and inspired me with their spirited, selfless dedication to the defence of freedom, equity, democracy, justice, and all the other noble qualities that make us human. I shared many seminars and workshops with Beatrix Tappeser, Christine von Weizsäcker, Elaine Ingham and Beth Burrows, who also provided a substantial amount of the material that appears in this book. Sue Mayer, Isobel Meister, David King, Ricarda Steinbrecher, Liz Hoskens, Helena Paul, Kristin Dawkins, Doug Parr, Philip Bereano, Jaan Suuricula and Jan Storms have kept me informed of their campaigns, and sent me valuable material besides, via the Internet.

This book is a tribute to the many public interest organisations from all over the world, as well as the official UN delegates from many countries. I would like to mention, in particular, Kalemani Joe Mulongoy, for his efforts in ensuring public participation in the democratic process while he was a member of the Secretariat of the Biological Diversity Convention in charge of Biotechnology and Biosafety, for insisting that all sides of the biotechnology debate should be heard by policy-makers while he was Director of the Programme on Biotechnology and Biosafety at the International Academy of the Environment in Geneva and, most of all, for his integrity and the sacrifice he has had to endure in the interest of bringing transparency and equity to the world.

Many thanks are due to Vandana Shiva, Brian Goodwin, Peter Schei, Richard Strohman, Harry Rubin, Jaan Suurkula, Joe Cumming, Edward Goldsmith, David Korten, Eric Schneider, Helena Paul, Christine von Weizsäcker, Giles Maynard, Alick Bartholomew, Busca Mileusnic, Steven Rose and, especially, Charles Jencks, for reading and commenting on earlier versions of the manuscript.

I would like to stress that this book is *not* a personal attack on genetic engineers or molecular geneticists. Among the latter, I wish to express my indebtedness to Sue Povey and Dallas Swallow, with whom I had ventured into molecular genetics in the early 1980s, and who taught me most of what I know about genetic engineering biosafety. Jeff Pollard and Ted Steele have taught me a lot about the fluid genome, and continue to keep me in touch with their work. Thanks are also due to Peter Lund, David Heaf and others of the *If*-gene group, with whom I have had many discussions on genetics and ethics.

Julian Haffegee not only assisted me in preparing this book, but kept the research in our laboratory going while I was fully occupied in writing. Christine Randall, my secretary, helped in organising my work and creating extra time for me in a hundred different ways. Last, but by no means least, I have benefited immensely from discussing many of the ideas in this book with Peter Saunders and Brian Goodwin, both of whom are involved in different aspects of the new science of organic complexity. Without their constant support and encouragement this book may well never have been finished.

Finally, I emphasise that none of those mentioned should be held responsible for the defects in my presentation, nor for the views expressed.

M.-W.H.
December, 1997

Contents

Chapter 1

The Unholy Alliance

Genetic engineering biotechnology is an unprecedented alliance between bad science and big business which will spell the end of humanity as we know it, and of the world at large. Genetic engineering biotechnology is inherently hazardous, but the genetic determinist mindset that misinforms both practitioners and the public takes hold of people's unconscious, making them act involuntarily, unquestioningly, to shape the world to the detriment of human beings and all other inhabitants.

Suddenly, the Brave New World Dawns

Suddenly, as the millennium draws to a close, men and women are waking up to the realisation that genetic engineering biotechnology is taking over every aspect of their daily lives. They have been caught unprepared for the avalanche of products arriving, or soon to arrive, in their supermarkets: rapeseed oil, soybean, maize, sugar beet, squash, cucumber.... It started as a mere trickle less than three years ago - first, the BST-milk from cows fed genetically engineered bovine growth hormone to boost milk yield, then the tomato, genetically engineered to prolong shelf-life. They provoked much debate and opposition; as did indeed, the genetic screening tests for an increasing number of diseases. Surely, we reasoned, we wouldn't, and shouldn't be rushed headlong into this brave new world....

Back then, in order to quell our anxiety, a series of highly publicised consensus conferences and public consultations were carried out. Committees were set up by many European Governments to consider the risks and the ethics, and the debates continued. The public was dimly aware of critics who deplored the idea of tampering with nature and scrambling the genetic code of species by introducing human genes into animals, and animal genes into vegetables. These critics warned of unexpected effects on agriculture and biodiversity, of the dangers of genetic pollution that could not be reversed. They warned of genetic discrimination and the return of eugenics as genetic screening and prenatal diagnosis became widely available. They condemned the immorality of the 'patents of life' - transgenic animals, plants and seeds taken freely from the Third World by geneticists of developed countries, as well as human genes and human cell lines from indigenous peoples.

But, by and large, the public was lulled into a false sense of security, in the belief that the best scientists in the country and the new breed of 'bioethicists' were busy considering the risks associated with the new biotechnology and the ethical issues raised. Simultaneously, glossy information pamphlets and reports were widely distributed by the biotech industries and their friends, and endorsed by government scientists. Their aim was to promote public understanding of genetic 'modification' - the term 'engineering' having long since been banished by the promoters on the grounds that it sounded too unfriendly and frightening. Genetic modification, we were told, was simply the latest in a seamless continuum of biotechnologies practised by human beings since the dawn of civilisation, from bread and wine-making to selective breeding. The significant advantage of genetic modification was that it was much more precise, as genes could be individually isolated and transferred as desired.

Thus, it was suggested, its possible benefits to humankind were limitless. There was something to satisfy everyone. For those morally concerned about inequality and human suffering, it promised to feed the hungry of the world by genetically modifying crops to resist pests and diseases and to increase yield. For those who despaired of the present global environmental deterioration, it promised to modify strains of bacteria and higher plants that could degrade toxic wastes or mop up heavy metals to clean up the environment. For those hankering after sustainable agriculture, it promised to develop greener, more environmentally friendly transgenic crops that would reduce the use of pesticides, herbicides and fertilisers.

That was not all. It was in the realm of human genetics that the real revolution would be wrought. It was dedicated to uncovering the entire genetic blueprint of the human being. This would eventually enable geneticists to diagnose, in advance, all the diseases that an individual would suffer in his or her lifetime, even before that individual was born, or even as the egg was fertilised *in vitro*. A whole gamut of specific drugs could then be designed to cure all diseases and, furthermore, to be tailored to individual genetic needs. The possibility of immortality was dangling from the horizons as the 'longevity gene' was isolated.

There were problems, of course, as there would be in any new technology. The ethical issues had to be decided by the public (by implication, the science was separate and not open to question.) The risks had to be minimised (again, by implication, the risks had nothing to do with the science.) After all, it was argued, nothing in life was without risk. One takes risks simply by crossing the road. The new biotechnology (i.e. genetic engineering biotechnology) was under very strict government regulation, and the government's scientists and other experts would see to it that neither the consumer nor the environment would be unduly harmed, wouldn't they?

Then came the relaxation of regulation on genetically modified products, on the grounds that over-regulation was compromising the 'competitiveness' of the industry, and that hundreds of field trials had demonstrated the new biotechnology to be safe. And, in any case, it was argued, there was no essential difference between transgenic plants produced by the new biotechnology and those produced by conventional breeding methods. (During a public debate with me, Henry Miller, a prominent spokesperson for the industry in the United States, even went so far as to refer to the varieties produced by conventional breeding methods, *retrospectively*, as "transgenics".[1]) This was followed, a year later, by an avalanche of products, approved or seeking approval for market, for which neither segregation from non-genetically engineered produce nor labelling was required. One was left to wonder why, if the products were as safe and wonderful as claimed, they could not be segregated, as organic produce has been segregated for years, enabling consumers to be given the choice of buying what they wanted.

Almost immediately, as though acting on cue, the Association of British Insurers announced that, in future, people applying for life policies would have to divulge the results of any genetic tests they had taken. This was seen by many as a definite move towards open genetic discrimination. A few days later, Ian Wilmut, a scientist at the Roslin Institute near Edinburgh, announced that they had successfully 'cloned' a sheep from a cell taken from the mammary gland of an adult animal. Dolly, the cloned lamb, was born. The popular media went wild with, at one extreme, heroic enthusiasm and, at the other, a Frankenstein-like horror. Of course it took nearly 300 trials to get one success, but no mention was made of the vast majority of the embryos that failed. Why, I wonder, has this work only come to public attention now, when the research has actually been going on for at least ten years?[2] Was that ethical? If it could be done in sheep, did it mean it could be done in other mammals? Are we now nearer to cloning human beings?

As this book was going to press, Chicago independent scientist Richard G. Seed announced (January 7) that he would begin work on cloning humans. President Bill Clinton reacted swiftly, calling for legislation to ban human cloning; as did thirteen European countries. Among those who have objected are the UK House of Commons Science and Technology Committee (STC) which has said it wants British law to be amended to ensure that human cloning is illegal, President Chirac of France and German Research Minister, Juergen Ruettgers,who have also called for an international ban on human cloning.

There is virtual unanimous consensus among doctors and scientists that the technique is "untested and unsafe and morally unacceptable". (note: Associated Press 1998). Seed scoffs at his detractors, saying that what is feared now will be tolerated and, eventually, enthusiastically endorsed.

A report appeared in the Norwegian daily newspaper *Dagbladet* the next day, side by side with an item on Dolly, headlined "Dolly is Eating Herself to Death". The article claimed that Dolly could not stop eating, even though she was more than twice the size of her litter mates. Professor Kjetill Jacobsen, a developmental biologist at Oslo University, was quoted as saying that the cloning process that created Dolly did not involve fertilisation of the eggs by the sperm, which may be crucial in establishing metabolic regulation

Isn't it time we put the brakes on before we are plunged headlong into the abyss?

Members of the public are totally unprepared, as they are plunged headlong, against their will, into this brave new genetically engineered world, in which giant, multinational faceless corporations will control every aspect of their lives, from the food they eat, to the baby they might conceive and give birth to.

But, isn't it a bit too late in the day, you may ask, to tell us that? Yes and no. Yes, because I, who should, perhaps, have known better, was caught unprepared like the rest. And no, because there are so many people who have been warning us of this eventuality, people who have campaigned tirelessly on our behalf, some of them going back to the earliest days of genetic engineering in the 1970s. Although we may have paid them little heed, no, it is not too late - if only because that is precisely what we are being encouraged to believe. A certain climate is created - that of being rapidly overtaken by events - that reinforces the feeling that the tidal wave of progress brought on by the new biotechnology is impossible to stem, so that we may be paralysed into accepting the inevitable. But, no, it is not too late, because we shall not give up. For the consequence of giving up will be the dawning of a brave new world, and, soon after that, maybe no world at all. The gene genie is fast getting out of control. The practitioners of genetic engineering biotechnology, regulators and critics alike, have *all* underestimated the risks involved, which are *inherent* to genetic engineering biotechnology, particularly when it is misguided by an outmoded and erroneous world view that comes from bad science. The dreams may already be turning into nightmares.

That is why people like me are calling for an immediate moratorium on further releases and marketing of genetically engineered products, and for an independent public inquiry to be set up to look into the risks and hazards involved, taking into account the most comprehensive scientific knowledge available, in addition to the social and moral implications. This would be most timely, as the evidence shows that public opposition to genetic engineering biotechnology has been gaining momentum throughout Europe and in the USA.

In Austria, a record 1.2 million citizens, representing 20% of the electorate, signed a people's petition to ban genetically engineered foods, as well as the deliberate release of genetically modified organisms and the patenting of life. Genetically modified foods were also earlier rejected by a lay people consultation in Norway, and by 95% of consumers in Germany, as revealed by a recent survey. The European Parliament voted by an overwhelming 407 to 2 majority to censure the Commission's authorisation of imports of Ciba-Geigy's transgenic maize into Europe, and to call for imports to be suspended while the authorisation was re-examined. The European Commission has decided that, in future, genetically engineered seeds will be labelled, and is also considering proposals for retroactive labelling. Commissioner Emma Bonino, is setting up a new scientific committee for genetically engineered foods, members of which are to be completely independent of the food industry. Meanwhile, European Commissioner on Agriculture, Franz Fischler, supports the complete segregation and labelling of production lines of genetically modified and non-genetically modified foods.

Like other critics before me,[3] I do not think there is a grand conspiracy afoot, though I do believe that there are many forces converging to a single terrible end. As the environmentalist, Susan George, has pointed out, "They don't have to conspire if they have the same world view, aspire to similar goals and take concerted steps to attain them."[4] I don't even think that 'they'- those who are actively pushing genetic engineering biotechnology, or merely being swept along in the process - are necessarily aspiring to any goals, nor taking rational concerted steps to attain them. Instead, they are simply converging involuntarily, like sleep-walkers, under the insidious influence of a certain world view, without conscious will, without conscious knowledge, towards the brink of oblivion.

Make no mistake, however. There is a pattern in the madness that has gripped the industrialised world at the end of the millennium. I shall show how the current global crisis of new and reemergent infectious diseases such as AIDS, ebola, cholera and tuberculosis, and of multiple drug-resistant strains of pathogens, are related to the reductionist scientific world view which, in turn, underpins both the prevailing genetic determinism that has dominated biology since Mendel and Darwin and the prevailing global economic system with its image of the 'economic man'. I am not out to *prove* a deterministic connection between economic and genetic reductionism. That would be to fall prey to the crude mechanistic framework that is best left behind. I am, however, concerned to reveal the links between global economics, biotechnology, runaway diseases and the possibility of impending genetic catastrophe, *unless we consciously will ourselves to break those links.*

Science is not Bad, but there is Bad Science

I said perhaps I should have known better. I am one of those scientists who have long been highly critical of the reductionist mainstream scientific world view, and have begun to work towards a radically different approach for understanding nature.[5] But I was unable, for a long time, to see how much science really matters in the affairs of the real world, not just in terms of practical inventions like genetic engineering, but in how a scientific world view can take hold of people's unconscious, leading them to act, involuntarily, unquestioningly, to shape the world to the detriment of themselves. I was so little aware of how science may be used, without conscious intent, to intimidate and control, to obfuscate, to exploit and oppress. I failed to recognise how that dominant world view generates a selective blindness to make scientists themselves ignore or misread scientific evidence. No wonder there is so much anti-science sentiment abroad.

The point, however, is not that *science* is bad - the charge that is too often made by the Green movement and by journalists in the popular media - but that there can be *bad science* that ill-serves humanity. Science can often be wrong. The history of science can just as well be written in terms of the mistakes it has made as in the series of triumphs with which it is usually credited. Science is nothing more and nothing less than a system of concepts for understanding nature and for obtaining reliable knowledge that enables us to live sustainably with nature. In that sense, one can ill afford to give up science, for it is through our proper understanding and knowledge of nature that we can live a satisfying life and ultimately learn to distinguish the good science which serves humanity from the bad science that does not. In this view, science is imbued with moral values from the start, and cannot be disentangled from them. Therefore it is bad science that purports to be 'neutral' and divorced from moral values, as much as it is bad science that ignores scientific evidence.

It is clear that I part company with perhaps a majority of my scientist colleagues in the mainstream who believe that science can never be wrong, although it can be misused. Or, like Professor Lewis Wolpert, who currently heads the Committee for the Public Understanding of Science in Britain, (thereby setting a really bad example), they carefully distinguish between science - neutral and value-free - and its application, technology - which can do harm or good.[6] This distinction between science and technology is spurious, especially in the case of an experimental science like genetics, and almost all of biology, where the techniques determine what sort of questions are asked and hence the range of answers that are important, significant and relevant to the science. Where would molecular genetics be without the tools that enable practitioners to recombine and manipulate genetic material from different sources? And, having manipulated the genetic material and noted the significant, triumphant results, it is then all too easy to see the world in genetic determinist terms - i.e. genes determine our destiny, and so, by manipulating our genes, we may also manipulate our destiny. It is an irresistibly heroic view - except that it is totally wrong and misguided.

It is also meaningless, therefore, to set up Ethical Committees which do not question the basic scientific assumptions behind the practice of genetic engineering biotechnology. Their brief is severely limited, often verging on the trivial and banal - such as whether a pork gene transferred to food plants might be counter to certain religious beliefs - in comparison to the much more fundamental questions involved, regarding eugenics, genetic discrimination and, indeed, whether gene transfers should be carried out at all. These committees can do nothing more than make the unacceptable acceptable to the public.

The debate on genetic engineering biotechnology is dogged by the artificial separation imposed between 'pure' science and the issues it gives rise to. Ethics are deemed to be socially determined, and therefore negotiable, while science is seen to be beyond reproach, as it follows the 'laws' of nature. The same goes for the distinction between science and technology - the application of science. Risk assessments are associated with the technology, leaving the science untouched. The technology, so the reasoning goes, may be bad for your health, but not the science. In this book I shall show why science cannot be separated from moral values nor from the technology that shapes our society. In other words, bad science is unquestionably bad for one's health and wellbeing, and should be avoided at all cost. Science is, above all, fallible and negotiable, because we have the choice to do or not to do. It should be negotiated for the public good. That is the only ethical position one can take with regard to science. Otherwise, we are in danger of turning science into the most fundamentalist of religions. And that, working hand in hand with corporate interests, will surely usher in the brave new world.

Bad Science and Big Business

What makes genetic engineering biotechnology dangerous, in the first instance, is that it is an unprecedented close alliance between two great powers that can make or break the world: science and commerce. Practically all established molecular geneticists have some direct or indirect connection with industry. This inevitably sets limits on what the scientists can and will do research on, not to mention the possibility of compromising their integrity as independent scientists.[7]

The worst aspect of this alliance is that it has been formed between science at its most reductionist and multinational monopolistic industry at its most aggressive and exploitative.

This was brought home to me during a one-week course I attended on Globalisation and Economics, taught by Martin Khor, a Cambridge-trained economist from Malaysia, who heads the Third World Network (TWN). The TWN, established in the 1980s, is a public interest organisation based in Penang, which has been in the front line of the struggle against the unequal relationships imposed by the industrialised North on the developing nations of the South. Working with Martin is quite an experience. He has such a talent for bringing out the best and the most in other people. Before you realise it, he has organised you, in the nicest way possible, to do in a day as much work as you may normally have done in a week. Not too many people suspect, therefore, that Martin has very impressive ideas of his own which he is able to put across to advantage with his panoramic grasp of world politics. Martin is one of those who, very early on, identified genetic engineering biotechnology as having a pivotal role to play in deepening the inequalities between North and South, as well as between the rich and poor everywhere else on earth. In the course of his lectures, it dawned on me that genetic engineering technology is really bad science working hand in glove with big business for quick profit, aided and abetted by our governments for the banal reason, as Martin says, that governments wish to be re-elected to remain in power.[8]

When I say 'bad science', I am not launching a personal attack on molecular geneticists themselves, many of whom may be skilled and conscientious scientists working within a flawed conceptual scientific framework. Speaking as a scientist who loves and believes in science, I have to say it is bad science that has let the world down and caused the major problems we now face, not the least of which is the promotion and legitimising of a particular world view. It is a reductionist, manipulative and exploitative world view. Reductionist because it sees the world as bits and pieces and denies there are organic wholes such as organisms, ecosystems, societies and communities of nations. Manipulative and exploitative because it regards nature and fellow human beings as so many objects to be manipulated and exploited for gain; life being a Darwinian struggle for survival of the fittest.

It is by no means coincidental that the economic theory currently dominating the world is rooted in that same *laissez-faire* capitalist ideology that gave rise to Darwinism. It acknowledges no values other than self-interest, competitiveness and the accumulation of wealth, at which the developed nations have been very successful. Already, according to the 1992 United Nations Development Programme (UNDP) Report, the richest fifth of the world's population has amassed 82.7% of the wealth, while the poorest fifth gets a piddling 1.4%. Or, put in another way, there are now 477 billionaires in the world whose combined assets are roughly equal to the combined annual incomes of the poorer half of humanity - 2.8 billion people.[9] Do we need to be more 'competitive' still to take from the poorest their remaining pittance? That is, in fact, what we are doing.

The governmental representatives of the superpowers are pushing for a globalised economy under trade agreements which erase all economic borders. "Together, the processes of deregulation and globalisation are undermining the power of both unions and governments and placing the power of global corporations and finance beyond the reach of public accountability."[10] The largest corporations continue to consolidate that power through mergers, acquisitions and strategic alliances. Multinational corporations now comprise 51 of the world's 100 largest economies, only 49 of which are nations. Many small biotech companies have been wiped out since the 1980s, and, by 1993, only 11 giant corporations were controlling agricultural biotechnology. These are now undergoing further mergers. The OECD (Organisation for Economic Cooperation and Development) member countries are at this moment working in secret in Paris on the Multilateral Agreements on Investment (MAI), which is written by and for corporations to prohibit any government from establishing performance or accountability standards for foreign investors. European Comissioner, Sir Leon Britten, is negotiating in the World Trade Organisation, on behalf of the European Community, to ensure that no barriers of any kind will remain in the South to dampen exploitation by the North and, at the same time, to protect the deeply unethical 'patents of life' through Trade Related Intellectual Property Rights (TRIPS) agreements.[11] So, in addition to gaining complete control of the food supply of the South through exclusive rights to genetically engineered seeds, the big food giants of the North can asset-strip the South's genetic *and* intellectual resources with impunity, up to and including genes and cell-lines of indigenous peoples.

There is no question that the mindset that leads to and validates genetic engineering is *genetic determinism* - the idea that organisms are determined by their genetic make-up, or the totality of their genes. Genetic determinism derives from the marriage of Darwinism and Mendelian genetics, of which I shall say more in later chapters. It basically believes that the major problems of the world can be solved simply by identifying and manipulating genes, for genes determine the characters of organisms. So by identifying a gene we can predict a desirable or undesirable trait, by changing a gene, we can change the trait and by transferring a gene, we can transfer the corresponding trait.

The Human Genome Project was inspired by the same genetic determinism that locates the 'blueprint' for constructing the human being in the human genome. It may have been a brilliant political move to capture research funds and, at the same time, to revive a flagging pharmaceutical industry, but its scientific content was suspect from the first. While I do not doubt that many individual geneticists working on the Human Genome Project are motivated by the prospect of pure scientific discovery, or of benefiting humanity, they have to realise that genetic discrimination and eugenics are both logical consequences of the ideology that provided the primary motivating force for the Project. Genetic engineering biotechnology promises to work for the benefit of humankind. The reality is something else.

- It is making products that, by and large, nobody needs and certainly not everybody wants but which are being forced on consumers anyway, by a lack of segregation and labelling.
- It displaces and marginalises all alternative approaches that address social and environmental causes of malnutrition and ill-health, such as poverty and unemployment, and the need for a sustainable agriculture that could regenerate the environment, guarantee long-term food security and at the same time, conserve indigenous biodiversity.
- It claims to solve problems that reductionist science and industry have created in the first place - widespread environmental deterio-

ration from the intensive, high input agriculture of the Green Revolution, and accumulation of toxic wastes from chemical industries. What's on offer now is more of the same, except with new problems attached.

- It leads to discriminatory and other unethical practices that are against the moral values of societies and community of nations.
- Worst of all, it is pushing a technology that is untried, and, *according to existing knowledge, is inherently hazardous to health and biodiversity.*

I am told that there is a tradition in social law for questioning inherently dangerous activities - that is, things that are not safe for citizens beyond a reasonable doubt. I shall make the case that genetic engineering biotechnology falls squarely into this category of activities. To proceed beyond this point is to subject the public to unacceptable risks. It is to break a fundamental, well-established, legal, ethical norm.

Let me enlarge on this last point here, as I believe it has been underestimated, if not entirely overlooked, by practitioners, regulators and many critics of genetic engineering biotechnology who have displayed a certain blindness to concrete scientific evidence in their conscious or unconscious commitment to an old, discredited paradigm. The most immediate hazards are likely to be in public health - which has already reached a global crisis, attesting to the failure of decades of reductionist medical practices - although the hazards to biodiversity will not be far behind.

Genetic Engineering Biotechnology is inherently Hazardous

According to the 1996 World Health Organisation Report, at least 30 new diseases including AIDS, ebola and hepatitis C, have emerged over the past 20 years, while old infectious diseases such as tuberculosis, cholera, malaria and diphtheria are coming back worldwide. Almost every month now in the UK, we hear reports on fresh outbreaks: *Streptococcus*, meningitis. *E. coli.* Practically all the pathogens are resistant to antibiotics, many to multiple antibiotics. Two strains of *E. coli* isolated in a transplant ward outside Cambridge in 1993 were found to be resistant to 21 out of 22 common antibiotics.[12] A strain of *Staphylococcus* isolated in Australia in 1990 was found to be resistant to 31 different drugs.[13] Infections with these and other strains will very soon become totally invulnerable to treatment. In fact, scientists in Japan have already isolated a strain of *Staphylococcus aureus* that is resistant even to the last resort antibiotic, vancomycin.[14]

Geneticists have now linked the emergence of pathogenic bacteria and of antibiotic resistance to *horizontal gene transfer* - that is, the transfer of genes to unrelated species by infection through viruses, through pieces of genetic material, DNA, by being taken up into cells from the environment, or by unusual mating taking place between unrelated species. For example, horizontal gene transfer and subsequent genetic recombination generated the bacterial strains responsible for the cholera outbreak in India in 1992,[15] and the *Streptococcus* epidemic in Tayside in 1993.[16] The *E.coli* 157 strain involved in recent outbreaks in Scotland is believed to have originated from horizontal gene transfer from the pathogen, *Shigella.*[17] Many unrelated bacterial pathogens, causing diseases from bubonic plague to tree blight, have been found to share an entire set of genes for invading cells, which have almost certainly spread by horizontal gene transfer.[18] Similarly, genes for antibiotic resistances have spread horizontally and recombined with one another to generate multiple antibiotic resistances throughout the bacterial populations.[19] Antibiotic-resistant genes spread readily by contact between human beings, and from bacteria inhabiting the gut of farm animals to those in human beings.[20] Multiple antibiotic-resistant strains of pathogens have been endemic in many hospitals for years.[21]

What is the connection between horizontal gene transfer and genetic engineering? Genetic engineering is a technology designed specifically to transfer genes horizontally between species that do not interbreed. It is designed to break down species barriers and, increasingly, to overcome the species' defence mechanisms which normally degrade or inactivate foreign genes.[22] For the purpose of manipulating, replicating and transferring genes, genetic engineers make use of recombined versions of precisely those genetic parasites causing diseases including cancers, and others that carry and spread virulence genes and antibiotic-resistant genes Thus the technology will contribute to an increase in the frequency of horizontal transfer of those genes that are responsible for virulence and antibiotic resistance, and allow them to recombine to generate new pathogens.

What is even more disturbing is that geneticists have now found evidence that the presence of antibiotics typically increases the frequency of horizontal gene transfer a hundred-fold or more, possibly because the antibiotic acts like a sex hormone for the bacteria, enhancing mating and the exchange of genes between unrelated species.[23] Thus, antibiotic resistance and multiple antibiotic resistance cannot be overcome simply by making new antibiotics, for *antibiotics create the very conditions that facilitate the spread of resistance.* The continuing profligate use of antibiotics in intensive farming and in medicine, in combination with the commercial-scale practice of genetic engineering, may already be major contributing factors for the accelerated spread of multiple antibiotic resistance among new and old pathogens that the WHO 1996 Report has identified within the past ten years. For example, there has been a dramatic rise, in terms of both incidence and severity, of cases of infections by *Salmonella*,[24] with some countries in Europe witnessing a staggering twenty-fold increase in incidence since 1980.

That is not all. One by one, those assumptions on which geneticists and regulatory committees have based their assessment of genetically engineered products as 'safe' have fallen by the wayside, especially in the light of evidence emerging within the past three to four years. However, there is still little indication that these new findings are being taken on board. On the contrary, regulatory bodies have succumbed to pressure from the industry to relax the already inadequate regulations. Let me list a few more of the relevant findings in genetics, which I shall deal with in more detail in the rest of the book.

We have been told that horizontal gene transfer is confined to bacteria. That is not so. It is now known to involve practically all species of animals, plants and fungi. It is possible for any gene in any species to spread to any other species, especially if the gene is carried on genetically engineered gene-transfer vectors. Transgenes and antibiotic-resistant marker genes from transgenic plants have been shown to end up in soil fungi and bacteria.[25] The microbial populations in the environment serve as the gene-transfer highway and reservoir, supporting the replication of the genes and allowing them to spread and recombine with other genes to generate new pathogens.[26]

We have been assured that 'crippled' laboratory strains of bacteria and viruses do not survive when released into the environment.

That is not true. There is now abundant evidence that they can either survive quite well and multiply, or they can go dormant and reappear after having acquired genes from other bacteria to enable them to multiply.[27] Bacteria cooperate much more than they compete. They share their most valuable assets for survival.

We have been told that DNA is easily broken down in the environment. Not so. DNA can remain in the environment where it can be picked up by bacteria and incorporated into their genome.[28] DNA is, in fact, one of the toughest molecules. Biochemists jumped with joy when they realised they no longer had to work with proteins, which lose their activity very readily. By contrast, DNA survives rigorous boiling. So when processed foods are approved on the grounds that there can be no DNA left in them, one should ask exactly how the processing is done, and whether the appropriate tests for the presence of DNA have been carried out.

The survival of 'crippled' laboratory strains of bacteria and viruses and the persistence of DNA in the environment are of particular relevance to the so-called 'contained' users producing transgenic pharmaceuticals, enzymes and food additives. 'Tolerated' releases and transgenic wastes from such users may already have released large amounts of transgenic bacteria and viruses, as well as DNA, into the environment since the early 1980s when commercial genetic engineering biotechnology began.

We are told that DNA is easily digested by enzymes in our gut. Not true. The DNA of a virus has been found to survive passage through the gut of mice. Furthermore, the DNA readily finds its way into the blood stream, and into all kinds of cells in the body.[29] Once inside the cell, the DNA may insert itself into the cell's genome and create all manner of genetic disturbances, including cancer.[30]

There are yet further findings pointing to the potential hazards of generating new disease-causing viruses by recombining artificial viral vectors or transgenic vaccines and other viruses in the environment. The viruses generated in this way will have increased host ranges, infecting and causing diseases in more than one species, and hence will be very difficult to eradicate. *We are already seeing such viruses emerging.*

- Monkeypox, a previously rare and potentially fatal virus caught from rodents, is spreading through central Zaire.[31] Between 1981-1986 only 37 cases were known. But there have been at least 163 cases in one eastern province of Zaire alone since July 1995. *For the first time, humans are transmitting the disease directly to one another.*

- An outbreak of hantavirus infection which hit southern Argentina in December 1996 was *the first in which the virus was transmitted from person to person.*[32] Previously, the virus was spread by breathing in the aerosols from rodent excrement or urine.
- New highly virulent strains of infectious bursal disease virus (IBDV) have spread rapidly throughout most of the poultry industry in the Northern hemisphere. They are now infecting Antarctic penguins and are suspected of causing mass mortality.[33]
- New strains of distemper and rabies viruses are spilling out from towns and villages to plague some of the world's rarest wild animals in Africa:[34] lions, panthers, wild dogs, giant otters.

None of this plethora of new findings has been taken on board by the regulatory bodies.[35] On the contrary, safety regulations have been relaxed. Members of the public are being used, against their will, as guinea pigs for genetically engineered products, while new viruses and bacterial pathogens may be being created by the technology every passing day.

The present situation is reminiscent of the development of nuclear energy which gave us the atom bomb and the nuclear power stations that we now know to be hazardous to health and environmentally unsustainable on account of the long-lasting radioactive wastes they produce. Joseph Rotblat, a British physicist who won the 1995 Nobel Prize after years of battling against nuclear weapons, has this to say, "My worry is that other advances in science may result in other means of mass destruction, maybe more readily available even than nuclear weapons. Genetic engineering is quite a possible area, because of these dreadful development that are taking place there."[36]

The large-scale release of transgenic organisms is much worse than nuclear weapons or radioactive nuclear wastes, as genes can replicate indefinitely, spread and recombine. There may yet be time to stop the dreams turning into nightmares if we act now, before the critical genetic melt-down is reached.

Notes

1. Head to head debate, Oxford Centre for Environment, Ethics and Society, Oxford University, Feb. 20, 1997.

2. "Scientists scorn sci-fi fears over sheep clone", *The Guardian*, Monday February 24, 1997 p. 7. Ian Wilmut, the senior scientist involved, was quoted as saying, "It will enable us to study genetic diseases for which there is now no cure and track down the mechanisms involved. The next step is to use the cells in culture in the lab and target genetic changes into that culture." In the same article, Lewis Wolpert, developmental biologist of University College London was reported as saying, "It's a pretty risky technique with lots of abnormalities." Also report and interview in the Eight O' Clock News, BBC Radio 4, Feb. 24, 1997.

3. As for instance, Spallone, 1992.

4. George, 1988, p.5

5. My colleague Peter Saunders and I began working on an alternative approach to neo-Darwinian evolutionary theory in the 1970s. Major collections of multi-author essays appeared in Ho and Saunders, 1984; Pollard, 1984; Ho and Fox, 1988.

6. Lewis Wolpert, who currently heads the Committee for the Public Understanding of Science argues strenuously for this 'fundamentalist' view of science. See Wolpert, 1996.

7. See Hubbard and Wald, 1993.

8. This was pointed out to me by Martin Khor, during a course on Globalization and Economics that he gave at Schumacher College, Feb. 3-10, 1997.

9. See Korten, 1997.

10. Korten, 1997, p.2

11. See Perlas, 1994; also WTO: New Setback for the South, *Third World Resurgence* issue 77/78, 1997, which contains many articles reporting on the WTO meeting held in December, 1996 in Singapore.

12. Brown *et al*, 1993

13. Udo and Grubb, 1990.

14. "Superbug spectre haunts Japan", Michael Day, *New Scientist* 3 May, p. 5, 1997.

15. See Bik, *et al*, 1995; Prager, *et al*, 1995; Reidl and Mekalanos, 1995.

16. Whatmore *et al*, 1994; Kapur *et al*, 1995; Schnitzler, *et al*, 1995; Upton *et al*, 1996.

17. Professor Hugh Pennington, on BBC Radio 4 News Feb. 1997. He has confirmed this to me in a personal interview on July 2, 1997.

18. Barinaga, 1996.

19. Reviewed by Davies, 1994.

20. Tschäpe, 1994.

21. See World Health Report, 1996; also Garrett, 1995, Chapter 13, for an excellent account of the history of antibiotic resistance in pathogens.

22. See Ho and Tappeser. 1997.

23. See Davies, 1994.

24. WHO Fact Sheet No. 139, January 1997.

25. Hoffman *et al*, 1994; Schluter *et al*, 1995.

26. See Ho, 1996a

27. Jager and Tappeser, 1996, have extensively reviewed the literature on the survival of bacteria and DNA released into different environments.

28. See Lorenz and Wackernagel, 1994.

29. See Schubbert, *et al*. 1994; also, *New Scientist* Jan. 4, p.24, 1997, featured a short report on recent findings of the group that were presented at the International Congress on Cell Biology in San Francisco, December 1996.30. Wahl *et al*, 1984; see also relevant entries in Kendrew, 1995, especially "slow transforming retroviruses" and "Transgenic technologies".

31. "Killer virus piles on the misery in Zaire" Debora MacKenzie, *New Scientist*, 19 April, p.12, 1997.

32. "Virus gets personal", *New Scientist* 26 April, p. 13, 1997.

33. "Poultry virus infection in Antarctic penguins", Heather Gardner, Knowles Kerry and Martin Riddle, *Nature* 387 (15 May), p. 245, 1997.

34. See Pain, 1997.

35. I first drew attention to the dangers of

horizontal gene transfer and related observations at a conference on food organised by the National Council of Women of Great Britain in April, 1996, and again, in a fully referenced paper presented at a Workshop on Capacity Building in Biosafety for Developing Countries organised by the Stockholm Environmental Institute in May, 1996 (see volume edited by Virgin and Frederick, 1996, for a shortened version of the paper), where I made it clear that capacity building in biosafety is urgently needed for *developed* countries. Subsequently, I sent the full paper with detailed references to the U.K. Ministry of Agriculture Fisheries and Food with lists of hazards and information gaps for risk assessment that I believe ought to be directly addressed by appropriate monitoring of field releases and specifically targeted research. I received a reply from the MAFF experts which stated that they could find little or no evidence for horizontal gene transfer by comparing gene sequences of organisms in existing databases. While dismissing some of the points I made as "highly unlikely", they did state that they were addressing many existing information gaps by funding on-going research. The reply indicates that risk assessments are indeed being done in the absence of much necessary basic knowledge. I replied to their experts, challenging their interpretation on most of the points, and have yet to receive any further response. I have also given my papers to representatives from the Department of the Environment, and have had no reactions from them at all.

36. Quoted in "The spectre of a human clone", *The Independent* Wednesday, 26 February, 1997. p.1.

Chapter 2

Genetic Engineering Biotechnology Now

The commercialisation of science in genetic engineering biotechnology has compromised the integrity of scientists, reduced organisms including human beings to commodities, intensified the exploitation and oppression of the Third World, and threatened human and animal health and biodiversity. It fuels the resurgence of eugenics and genetic discrimination against non-white populations, minority groups and all the politically dispossessed peoples of the world. It results in a monolithic wasteland of genetic determinist mentality that is the beginning of the brave new world.

Genetic Engineering Then and Now

Genetic engineering is a set of techniques for isolating, modifying, multiplying and recombining genes from different organisms. It enables geneticists to transfer genes between species belonging to different kingdoms that would have no probability of interbreeding in nature. So, for example, a fish gene can be transferred to a tomato, human genes can be transferred to sheep, pigs or to the bacterium *E. coli*, that inhabits the gut of all mammals.

Genetic engineering originated in the 1970s, as the result of the discovery of several key techniques in molecular genetics (see Chapter 3). Soon afterwards, the molecular geneticists who discovered the techniques, or were in the forefront of developing and using genetic engineering, became aware of the dangers of opening a Pandora's box. They saw the distinct possibility of inadvertently, or intentionally, creating pathogenic strains of viruses or bacteria by recombining genes in the laboratory. This led to the Asilomar Declaration,[1] which called for a moratorium on genetic engineering until appropriate regulatory guidelines had been put in place. The scientists were acting responsibly. They were the first to recognise the dangers, so they brought the matter to public attention and, at the same time, imposed a moratorium on their own research.

I had the experience of setting up a genetic engineering laboratory in my university in the 1980s, when experimenters were taught to take great care to ensure that the microorganisms and genes we were manipulating remained contained in the laboratory. We had to work in special flow hoods, nothing got flushed down the sink and accidental spills had to be wiped down immediately and thoroughly with antiseptics. Wastes were autoclaved before disposal, or incinerated. There was no question that we were to think it safe to release genetically engineered organisms into the environment, even though the strains we dealt with were genetically crippled or handicapped in some way, so that they would not be expected to survive when accidentally released. (In fact, that assumption has now been proved wrong, so it was just as well that we operated on the precautionary principle).

I went into genetic engineering as someone interested in evolution, to learn about genetic processes that could respond, in a repeatable, and non-random way, to the environment, and hence influence the course of development and evolution. Basically, I did not accept the conventional, simplistic view which says that evolution occurs mainly by the natural selection of random genetic mutations. I left the field in the late 1980s, satisfied that genetic determinism had been completely invalidated by the findings of the new molecular genetics which indicated, among other things, that genes and genomes could respond in non-random ways to the environment. This left me free to begin researching into the biophysics of self-organisation, of how it is that organisms function as coherent wholes, and not just as collections of genes.[2] Unfortunately, well into the 1990s, the mindset of mainstream science hasn't changed at all. If anything, it has got considerably worse. Genetic determinism is rife, not least because it is perfect for promoting genetic engineering biotechnology.

Now, in the 1990s, the risks from genetic manipulation have become far greater. Genetic engineering techniques are ten times faster and more powerful. The new breed of genetically engineered organisms (or transgenics) which are deliberately released on a large scale are designed to be ecologically vigorous and, therefore, potentially much more hazardous than the genetically crippled microorganisms which were engineered for contained use in the laboratory in the 1970s.[3] Where is the voice of science now? The scientists say it is for the politicians and the public to decide. Of course, the public should decide, but that does not absolve scientists from their special responsibility as both citizens *and* scientists. As C.P. Snow writes on the scientists' responsibility with regard to making the atom bomb, " ..It is not enough to say that scientists have a responsibility as citizens. They have a much greater one than that, and one different in kind. For scientists have a moral imperative to say what they know...."[4]

Although an increasing number of scientists are critical of genetic engineering biotechnology, there has been no equivalent of the Asilomar Declaration from molecular geneticists in the 1990s calling for a moratorium. As Harvard biologist Ruth Hubbard,[5] notes, many of the current top molecular geneticists either own biotech companies or are collaborating with, or working for, such companies. Genetic engineering biotechnology is the commercialisation of science on an unprecedented scale.

"Scientists are increasingly being forced to get into bed with big business...Where research was once mostly neutral, it now has an array of paymasters to please. In place of impartiality, research results are being discreetly managed and massaged, or even locked away if they don't serve the right interests...More pernicious, according to *The Sci Files*, a new BBC TV series...is the slide into self-censorship in an attempt to ensure that the contracts keep coming."[6] We have moved far away, indeed, from the world of C.P. Snow.

The change is partly out of necessity as, over the past decade, the British Government has cut more than £1 billion from research funding, and partly because there's money to be made. Even what remains of government research support is strongly committed to a closer link with industry, as was made clear by a 1993 science white paper. It stressed the need to concentrate on research that would help the economy, and genetic engineering biotechnology is clearly seen to be the prime candidate.

"The Government is committed to ensuring that the United Kingdom can remain a leading developer and producer of biotechnology ...World markets for biotechnology products have been estimated to exceed £70 billion by the year 2000, growing at some 30% at year, and as one of the most active European countries in the field, the United Kingdom's share of sales dependent on biotechnology could rise from £4 billion to £9 billion by 2000..."[7]

The 'Patenting of Life'

The commercialisation of genetic engineering has been growing steadily since the 1970s. The first corporation, Genentech, was formed - even as a moratorium was being debated in 1976 - by molecular geneticist Paul Berg, who had signed the Asilomar Declaration a year earlier. The next milestone was the 1980 U.S. Supreme Court ruling that genetically engineered microorganisms could be patented. Then came the $3 billion U.S. federally-funded Human Genome Initiative,[8] which opened the floodgate to 'patents on life'. A long list of patents have already been granted, and many more are pending, on controversial 'inventions' such as:

- transgenic organisms, human genes and gene fragments
- a human cell line established from the spleen of a patient removed as part of cancer therapy[9]
- cell lines from indigenous tribes obtained - without informed consent - ostensibly for the study of human diversity
- seeds and plant varieties taken by Northern 'bio-prospectors' from indigenous communities in the Third World who freely provided the material as well as their knowledge.

These patents go to feed the mushrooming biotech industry, greedy for products and quick profit.

To facilitate patenting for commercial exploitation, the Trade-Related Intellectual Property Rights (TRIPS) treaty was introduced in the Draft Final Act of the General Agreement on Tariffs and Trade (GATT), before the latter was dissolved and replaced by the World Trade Organisation (WTO). The TRIPS treaty "effectively excludes all kinds of knowledge, ideas and innovations [for patenting] - that take place in the 'intellectual common' - in villages among farmers, in forests among tribals."[10] It regards as invention solely those acts carried out within the framework of western science. Science is here working hand-in-hand with corporate interests to define what is scientific and what is not, and hence what qualifies as a real invention for the purpose of financial reward. One should not underestimate the power of science to legitimise and exclude and hence to exploit and oppress.

Patenting plant varieties from Third World countries robs indigenous farmers of their livelihood, and can have widespread repercussions. The neem plant in India, for example, whose seed oil possesses insecticidal and many medicinal properties, has been freely available for millennia, so much so that the health care system of the whole of India is dependent upon it. As soon as it was 'discovered' and patented by the U.S. company, W.R. Grace, it became a scarce commodity. Its market value shot up a hundred-fold within two years, to put it well beyond the means of most ordinary people. A national health system has thereby been seriously undermined.

The intellectual property right over genetic resources is emerging as a major North-South issue. It began with the enactment of an International Convention - the Union for the Protection of New Varieties of Plant (UPOV) - in the early 1960s, which gave property rights to plant breeders for varieties improved through human intervention. The source material, obtained freely from the biodiverse countries of the South, was considered the 'common heritage of mankind' and hence not subject to private ownership. This gave free access to corporate interests to bioprospect in the South, and started the process of the increasingly arbitrary categorising of 'innovation' by Northern companies, while the real innovative contributions of local communities were denied as such.[11] Under this unfair convention, Northern countries are allowed to take freely from the South, as 'common heritage', genetic resources which are then returned to them as priced commodities. Strong protest from Third World countries led to a meeting in 1987 of the Food and Agricultural Organisation Commission on Plant Genetic Resources which recognised the contribution of traditional farmers in developing the plant, but that right was not vested in individual farmers. Instead, it accrued to the farmers' governments to receive assistance in maintaining the genetic resources. In other words, the North is 'obliged' to help the South, tied into the concept of aid and dependency that has for centuries allowed the North to exploit the South. An international gene fund was set up to establish the farmers' rights, but the lack of contributions from Northern corporations and their governments made this fund inoperative. The TRIPS proposal is generally seen as the latest attempt to formalise the continuing piracy of Third World genetic resources by Northern biotech companies, effectively sanctioned by the science of genetic engineering.

The United Nations Convention on Biological Diversity (CBD), signed in Rio de Janeiro in June, 1992, is intended to conserve biological diversity in an equitable way. As the major genetic diversity is centred in the poorer, less developed countries of the world, and this is the material genetic engineering wishes to use, the Convention will play an important role in determining the socio-economic effects of the technology. Dr Tewolde Egziagher, an agronomist trained at Cambridge University and a UN delegate from Ethiopia, has emerged as a key spokesperson for the entire African group of nations. His speeches are occasions in themselves, always delivered with great eloquence, full of gentle ironies and warmth, inspiring solidarity and consensus while, at the same time, pinpointing clearly the source of contention. Tewolde works closely with the TWN and I got to know him well. He was one of the prime movers for putting the International Biosafety Protocol on the agenda of the CBD as it was signed in 1992, and has been responsible for major inputs to the Protocol ever since. He has always insisted that the socio-economic impacts of genetic engineering biotechnology should be included as part of the biosafety risk assessment. One of the main reasons is that the CBD has done nothing to stem the drain of natural resources of the South to the North. On the contrary, the very basis of sustainability and long-term food security of the South is now threatened as the result of genetic engineering agricultural biotechnology promoted under the Convention.[12]

The CBD has been hailed as "the culmination of two decades of arduous international efforts in which the conservation of biological diversity is being recognised as a common concern of humankind, and considered an integral part of the development process.....[It would] reconcile the need for conservation with the concerns for development based on justice and equity".[13] Instead, pilfering of biological diversity has intensified as agricultural bio-technology drives 'gene-hunters' to prospect for commercially lucrative genetic resources in the South, in the new regime of intellectual property rights that allows patenting of living organisms and their genes.

To make things worse, large proportions of the biological diversity of the South are already held in 'gene banks' as *ex situ* collections, and the North is insisting that such *ex situ* collections should be excluded from the Convention, with the result that they will be freely available for exploitation by biotech interests. While negotiations are still going on, European botanical gardens and other collections of tropical plants have already been approached by a representative of Phytera, a U.S. biotech company specialising in pharmaceuticals, to gain access to items in their collections. A small initial fee would be paid on delivery of each item and, in the case of successful market introduction of a product derived from the item, a small percentage of the profit would be returned to the garden or herbarium.[14]

Life as Commodity

The strongest objection to the 'patenting of life', however, is that it has turned organisms, including parts of human beings, into saleable commodities. This is morally repugnant, especially to many indigenous cultures in the Third World, and has also united diverse groups in the North. These include environmental activists and religious organisations as well as ordinary citizens who feel that the final frontier of human decency has been breached in the name of free enterprise. But this is merely the logical convergence of the instrumental view of nature sanctioned by reductionist science and its kindred capitalist ideology that is driving the new biotech industry towards the limit of the exploitable.

As a result of world-wide opposition, a number of patents granted have been revoked or retracted by the claimants.[15] For example, at the end of 1995, the U.S. Center for Disease Control dropped its patent claim on a cell line from a Guyami woman from Panama after protests from the World Council of Indigenous Peoples and the Guyami General Congress. At the same time, UNESCO's international bioethics committee, rather than endorsing the Human Genome Diversity Project (part of the Human Genome Project set up ostensibly to document and preserve genes from indigenous peoples before they become extinct) endorsed the criticisms of the project raised by indigenous peoples and their governments. In 1994, The U.S. Patent and Trademark Office provisionally revoked Agracetus' patent on all genetically engineered cotton, and the Indian government revoked the same company's application for a patent in India of genetically engineered cotton. Many legal oppositions were filed in the European Patent Office against the patent of the 'oncomouse' - a transgenic mouse designed to be prone to cancer. That has effectively blocked more than 300 other applications for patents on animals, pending the outcome of the oncomouse case.

The European Parliament, responding to public opposition, voted to reject the Draft Directive on the Protection of Biotechnological Investigations in its first round in March 1995. By the end of 1995, the European commission published a revised Draft Directive, though the changes are cosmetic and the wording ambiguous, to say the least. It allows the patenting of any microorganisms, any plant or animal derived from 'a microbiological process', and any isolated human gene sequence, of whatever known or unknown function. In short, it would give biotech companies patent protection for products of genetic engineering over and above any other real invention covered by existing patent laws. As such, it requires no act of invention. This will grant unprecedented monopolistic rights to corporate patent holders, which not only effectively prevent competitors from developing related products, but also researchers from carrying out scientific research. As a group of scientists writing in the correspondence pages of *Nature* point out,

"Advances in biotechnology are already patentable under European patent law. What is at issue is whether these patents should be very much broader in scope than those in other fields, and in particular, whether someone who isolates and characterises natural material should be able to patent not just the method by which this was done but also the material itself. If this principle had been applied in chemistry, the elements would have been patented, and indeed, the directive does refer to 'elements of plants and animals'."[16]

A controversial recent patent, granted in both the USA and in Europe to the company Biocyte, covers all human blood cells gained from the umbilical cord of new-born babies.[17] These cells are routinely used for therapeutic bone-marrow transplant, without cost, and in the time-honoured tradition of offering the gift of life to those in need. In future, no such therapeutic transplantation nor research will be able to be carried out using such cells unless licence fees are paid to the company concerned.

The political repercussions of intellectual property rights agreements on living organisms are far-reaching.[18] In April, 1997, a letter was sent by the U.S. State Department to the Royal Thai Government regarding the latter's draft legislation allowing Thai doctors to register traditional medicines. It states that "Washington believes that such a registration system could constitute a possible violation of TRIPS and hamper medical research into these compounds", despite the fact that Thailand is not obliged to comply with TRIPS until at least the year 2000, and medical practices may be exempted. The U.S. has already threatened the Ecuadorian Government with the cancellation of trade preferences if the latter does not ratify a bilateral agreement on intellectual property rights, and Ethiopia, Panama and Paraguay have been added to the list of countries which limit U.S. commercial interests. The U.S. unilaterally cancelled half of Argentina's trade benefits valued at $260 million on the grounds that Argentina's intellectual property laws did not comply with "international standards". In June, 1997, the U.S. Ambassador to India announced that "certain areas of research and training will be closed to cooperation" if India fails to amend its patent laws, threatening some 130 scientific projects currently supported by the U.S.-India fund. The U.S. has also filed formal complaints with the WTO against India and Pakistan regarding their national patent laws governing pharmaceutical and agricultural chemical products. The same complaints have been filed against the Danish Government.

On July 16, the European Parliament succumbed to commercial pressures and voted to accept the revised Directive on patenting.

The Question of 'Safety'

The Convention on Biological Diversity Agenda 21 contains a chapter (no.16), entitled, "Environmentally Sound Management of Biotechnology", which recommends that some billions of dollars of the UN budget be committed to genetic engineering biotechnology so as to increase food yield to feed the hungry, to improve human health and control population, to purify water, to clean up the environment, reforest wasteland, in short, to solve all the problems of the Third World. At the same time, the hazards, as well as impacts from the new technology, are consistently glossed over by the use of some of the verbal ploys already mentioned: for instance, the supposed continuity between conventional biotechnology (like wine-making) and modern biotechnology, i.e., genetic engineering; the claim that years of experience have demonstrated modern biotechnology to be safe, and the substitution of the less emotive term 'genetic modification' for 'genetic engineering'. It is remarkable how much of the same propaganda material eventually ends up in all the literature promoting 'public understanding' of genetic engineering biotechnology, and is echoed by all the major spokespersons of the industry, including many scientists who should know better.[19]) This comes at a time when no other UN project is being funded under the Commission for Sustainable Development. Chapter 16 of the Convention is generally regarded as a thinly veiled attempt to promote and subsidise the biotech industry. This would not be surprising, given the openly partisan view expressed by the UK Government in favour of the industry, which is not atypical of the views of other industrialised countries in the North, with a few notable exceptions such as Austria, Denmark, Norway and Sweden.

Moreover, as opposition to genetic engineering biotechnology has been gathering momentum in developed countries, the industry is targeting the Third World for test-sites as well as markets. Critics are justifiably concerned about the uncontrolled releases of transgenic organisms in the Third World, and about people being used as human guinea-pigs for testing genetically engineered drugs and vaccines. By 1994, there had already been at least 90 releases of transgenic crops in non-OECD countries and Mexico, a third of which were by multinational corporations such as the U.S. based Monsanto and Calgene, and the Swiss company, Ciba Geigy.[20] A rabies vaccine containing a live virus was tested on cattle in Argentina without authorisation, and farm-workers who were not informed of the experiment were subsequently found to be infected with the virus.[21]

Most Third World countries have neither the legal framework nor the capacity to regulate genetic engineering. The same is true, however, of developed countries. There is, at present, no legal control over genetically engineered versions of drugs and chemicals already approved for the market, nor is there any legal requirement that they be labelled as such. The same goes for transgenic crop-plants and other products, where 'substantial equivalence' is claimed (see Chapter 1). As there is no clear definition as to what constitutes 'substantial equivalence', it leaves a gaping loophole in safety assessment.

There have already been serious indications of what can happen when safety is ignored. 'Unexpected' toxins and allergens have been associated with genetically engineered foods. The first case was in 1989, when trace contaminants in the amino acid tryptophan, produced by a Japanese biotech company using a newly genetically-engineered microorganism, were implicated in an outbreak of a mysterious illness, eosinophilia-myalgia syndrome (EMS), which led to 37 deaths and more than 1,500 affected.[22] More recently, a soybean genetically engineered with a brazil nut gene was found to be allergenic to people sensitive to brazil nuts,[23] while a strain of yeast, engineered to ferment faster, was found to accumulate a metabolite at mutagenic levels.[24] As portents of the ecological hazards of releasing transgenic crops, field trials have shown that herbicide resistance in transgenic potato[25] and transgenic oilseed rape[26] have spread to weed relatives within a single growing season, thereby creating herbicide-resistant super-weeds. A genetically engineered soil bacterium, thought to be quite harmless, turned out drastically to inhibit the growth of wheat seedlings.[27]

Thus, by 1995, there were widespread and justified concerns about the health hazards and ecological impacts, as well as the socio-economic implications of genetic engineering biotechnology which would arise from the erosion of farmers' rights - displacing small local farmers - or from the substitution of indigenous crops by transgenic organisms grown in the laboratory. This convinced all Third World countries (the G7 and China), Eastern European and most Western European countries that a legally-binding International Biosafety Protocol for the handling and transfer of genetically engineered organisms should be established as a matter of urgency. This has been openly opposed by the United States (which has, as yet, failed to rectify the Biodiversity Convention, on the grounds that it would reduce U.S. 'competitiveness'). The U.S. is supported by the UK, Australia, Germany, the Netherlands, and representatives of the biotech industry, who have consistently rejected a legally binding biosafety protocol, calling instead, for 'voluntary guidelines'.

At the time that I became involved in biosafety issues, the list of hazards from the products of genetic engineering biotechnology was growing, but the official UN panel of biosafety experts remained unmoved. Its report in May 1995 still maintained that there was no difference between conventional and modern biotechnology, and that years of experience had shown modern biotechnology to be safe. It even proposed a relaxation of the voluntary guidelines originally drawn up by the UK and the Netherlands, and already considered inadequate by many scientists. The TWN, represented by Martin Khor, Chee Yoke Ling and Nijar Gurdial, together with the Edmonds Institute in the United States (run single-handedly, by the incredible Beth Burrows) began bringing scientists to UN conferences. These prime activists were kept busy organising workshops and seminars during the lunch-hours, issuing briefing papers and daily bulletins ('scandal sheets') which Beth Burrows and members of other non-government organisations sometimes stayed up all night to produce. These briefing papers and daily bulletins were eagerly awaited and snapped up by the delegates, especially those from Third World countries.

The seminars and workshops were a significant part of the process. One regular member of our group was Christine von Weizsäcker of Ecoropa. She has the advantage of having been trained both in biology and philosophy, and always added an extra dimension to our seminars which were typically interdisciplinary, dealing with all aspects of biosafety: the scientific, the legal and the socio-economic. You will be meeting other members of our 'independent panel' in due course in this book, but Vandana Shiva, already well-known for her writings and political activism, is, without doubt, the star speaker. Not only is Vandana well-versed in the scientific details of genetic engineering, which she has picked up in no time at all, but she is incredibly well-informed on all the relevant issues of the day. A powerful and eloquent speaker, sharp-witted and a top debater to boot, she is an absolute terror to encounter in opposition. It is a pleasure to share a platform with her, if only to watch the other side cower.

Seminars and workshops were not enough. Martin Khor organised a group of us, scientists and legal experts from many countries, both North and South, to draft an alternative, independent report on biosafety, based on the most up-to-date scientific findings, spelling out the risks and calling for tighter monitoring and control. (In retrospect, we should have called for a moratorium there and then.) This document was published by the TWN and circulated to the UN delegates at every opportunity.

Our legal experts include Chee Yoke Ling and Nijar Gurdial. Yoke Ling is the youngest of the group. She trained in law at Cambridge University but gave up the possibility of a lucrative practice in Hong Kong to work for the TWN. Yoke Ling is deceptively frail, slim, good-looking and good-natured, which often led opponents to underestimate her - much to their regret afterwards. Nijar Gurdial *did* give up his practice to work for the TWN, despite having a serious heart condition. He recognised the implications of the new genetics for biosafety almost immediately and worked them into our Biosafety Report. Gurdial and Yoke Ling are both great fun to be with (Gurdial told so many jokes one evening at dinner that we were thrown out of the restaurant for disorderly behaviour.) They have both been involved in intellectual property rights and biosafety issues since 1992. Yoke Ling, in particular, knows quite a bit about the history of biosafety and what goes on behind the scenes.

The biotech industry's vehement opposition to a Biosafety Protocol was a major factor preventing the Bush Administration from signing on to the Convention, and even when the Clinton Administration signed on, ratification was refused by the U.S. Congress. When a Protocol was finally agreed by an overwhelming majority of the countries in July 1995, in Madrid, the biotech industry, together with the Clinton Administration, began to work hard to undermine its effectiveness. In particular, they insisted on excluding liabilities and compensation, and socio-economic impacts of genetic engineering biotechnology. Experience with the Green Revolution has taught many countries in the Third World to be, rightly, wary of the new biotechnology, with its possible impacts on the environment, agricultural and natural biodiversity, as well as the livelihoods of small farmers.

Prior to a subsequent meeting on biosafety in Aarhus, Denmark, in July 1996, the U.S. Clinton Administration sent messages, via its Embassies in developing countries, to relevant ministries asking for their position on biosafety. The messages carried underlying threats that a 'badly drafted protocol' would deny developing countries the benefits of genetic engineering and might even be illegal under the trading rules of the World Trade Organisation.[31]

Science is thereby enmeshed in the tangle of trade, technology, perceived risks, and lack thereof, and international intrigue. It provides the ideological backdrop, the material substance and the terms in which an exploitative, unequal trade relationship can be defined between the North and the South. The North is given sole recognition of its science, in the concrete form of intellectual property rights, while the South is deprived not only of its own sciences, but of its right to exclude the encroaching science of the North and the North's usurpation of the sciences of the South as its own in widespread acts of 'bio-piracy'.

The same science claims to override any possible objections from the European Union to imports of genetically engineered foods and any requirement for segregation and labelling. In June, 1997, the biotech industries wrote to President Clinton in preparation for the up-coming G8 meeting,[32] "Because trade is so important to American agriculture and the U.S. food industry, it is imperative that policy and regulations governing international commerce of genetically modified food and agricultural products are based on sound science and not just emotion which often turns into pure hyperbole.

It is also important to note that segregation of bulk commodities is not scientifically justified and is economically unrealistic.

"Some officials of the EU advocate requirements that could be considered non-tariff trade barriers to the U.S. and other countries exporting to the EU. It is critical the EU understand at the highest level that the U.S. would consider any trade barrier of genetically modified agricultural products, be it discriminatory labelling or segregation, unacceptable and subject to challenge in the World Trade Organisation (WTO)."

The U.S. Government took a correspondingly firm stand. Agriculture Secretary Dan Glickman told a conference of the 44-nation International Grains Council in London that it would not tolerate segregation or labelling,

"Sound science ought to be the only arbiter," he said. "The greatest threat to free trade is bad and phoney science....We know that biotechnology holds out our greatest hope of dramatically increasing yields."

That makes it all too clear that 'sound science' is that which sanctions free trade and protects biotech interests at the expense of any safety considerations. There is a 'science war' going on in the real world which is far, far more important than that going on in academia between the 'relativists' in sociology - who say science is a social construct - and the fundamental 'absolutists' of the scientific mainstream - including Lewis Wolpert and others - who regard science as the ineluctable, eternal laws of nature. They are both wrong and irrelevant. *This* science war is for real. It is what makes former ivory-tower academics like myself take to the political streets.

Official Disinformation on Safety

The official position of the U.S. on biosafety comes from a U.S. National Research Council report, *Field Testing Genetically Modified Organisms: Framework for Decisions*, which states *a priori* that "..no conceptual distinction exists between genetic modification of plants and microorganisms by classical methods or by molecular techniques that modify DNA and transfer genes." Similar statements have made their way to Chapter 16 of Agenda 21 mentioned above, and many other UN documents on biosafety. It is no coincidence that is also the position adopted by the biotech industry.

This statement is obviously untrue, as I shall show in the next chapter. But, the *a priori* assumption that there is no difference between genetically engineered varieties and those made by traditional breeding methods has meant that field tests are both inadequately designed and inadequately monitored for safety. It is on the basis of such inadequate field tests that transgenic crops have been approved as safe for human and animal consumption without any legal requirement for appropriate tests for safety to be carried out.

In 1995, scientists from the U.S. Government's Environmental Protection Agency issued a report charging their own agency with failing to assess the risks associated with the massive release of a new living organism that could not be contained or eradicated.[33] The organism in question was a *Rhizobium* - a bacterium which normally lives as a nitrogen-fixing symbiont in the roots of legumes - engineered with, among other things, an antibiotic-resistant marker gene from the pathogen, *Shigella*, which causes dysentery and infantile gastro-enteritis in humans. Among the risks not assessed were the potential transfer of antibiotic resistance to other pathogens and the subsequent creation of drug-resistant diseases in humans, livestock and wildlife, toxicity to humans, and hazards to the ecological environment. Incidentally, no data were produced as to whether the genetically engineered *Rhizobium* was effective in improving the yield of crop-plants.

The Union of Concerned Scientists in the USA have evaluated the data on field trials to see whether they support the conclusion of safety. Margaret Mellon and Jane Rissler have this to say, "...care should be taken in citing the field test record as strong evidence for the safety of genetically engineered crops. It is not. Unless they are redesigned to collect environmental data, the field tests do not provide a track record of safety, but a case of 'don't look, don't find'."[34] To that, Henry Miller, (that same staunch defender of genetic engineering biotechnology mentioned in Chapter 1), retorted that, as there was no essential difference between genetically engineered organisms and the strains obtained by conventional methods, field trials were done on a 'don't need, don't look' basis.[35] That is a strong admission that the scientific framework is used to legitimise the culpable lack of adherence to regulation on the part of the scientists, as well as the lack of actual regulation imposed by the regulatory bodies. The biotech industry in the U.S. is now asking for even less regulation than currently exists for traditional approaches.[36]

I have examined Monsanto's application to release transgenic sugar beet in the UK between 1997 and 1999, which was sent to me, on request, from the Department of the Environment. Crucial information is not available to the public, as it is deemed to be 'Confidential Business Information', and no actual scientific data are supplied throughout the application, only a series of unsupported assertions and statements. I was no wiser afterwards as to the nature of the transgenic organism to be released. In addition, it confirmed my worst fears of the inadequacy of the information required for a proper risk assessment. I have written to the Department of the Environment of my concerns,[37] and have since received a very inadequate reply, which fails to address most of the points I raised.

There is, at present, no evidence to support the assertions of the biotech industry and the regulatory bodies that genetic engineering biotechnology and its products are safe. Tests for toxicity and allergenicity of food products, where they have been carried out at all, are solely targeted at *known* allergens and toxins, and not designed to reveal unexpected products resulting from the genetic engineering.[38] Not only are field trials inadequately designed and monitored, the regulation is perfunctory. Horizontal gene transfer has never included monitoring for field tests. The departments concerned are not equipped to cope with the flood of applications for approval. Yet, even those weak regulations have been further relaxed.

From the beginning, the scope of risk assessment was restricted. It did not take account of all scientific findings - the absence of evidence was too often taken to be evidence of absence. Moreover, risk assessment takes place in a social vacuum. Socio-economic impacts are consistently excluded from consideration. At such times, expert panel members insist that they will only consider scientific evidence, while it is clear that they do not consider scientific evidence at all. Once again, science is being used to legitimise and exclude. But let us look at some of the socio-economic impacts that are already in evidence.

The Human Genomania[39]

The title of this section is taken from a paper written by Professor Ruth Hubbard, a politically concerned scientist, leading feminist and a critic of genetic determinism since the 1970s when genetic engineering began. Ruth points out that, even before the Human

Genome Project got underway, the diagnosis of genetic disease such as sickle cell anaemia, which predominantly affects Afro-Caribbeans, has resulted in individuals being discriminated against in health insurance and in employment within the U.S. The diagnosis of diseases for which no cure is forthcoming is of questionable value, as even for many so-called 'single-gene' diseases, the clinical prognosis can vary widely from individual to individual, simply because genes do not function in isolation from all other genes, but will differ considerably from individual to individual.

Nonetheless, geneticists are now attempting to identify genetic 'predispositions' and 'genetic propensities' for disorders such as cancer, diabetes, schizophrenia and, worse, for conditions such as alcoholism, homosexuality and criminality that, overwhelmingly, come under the influence of environmental and social factors. This not only diverts attention from the real causes, but also stigmatises individuals, through placing the blame of society's ills on people's genes,[40] and through the arbitrary categorisation of the 'normal' versus the 'abnormal'. The International League of Societies of Mentally Handicapped Persons gave evidence to UNESCO's International Bioethics Committee, pointing to the invisible social, legal and financial pressures already forcing women to abort disabled foetuses, and to the fact that genome research could 'geneticise' social policies and reduce financial support for disabled people.[41] So-called 'therapeutic' abortions of affected foetuses and the contemplation of germline gene therapy and genetic manipulation are, respectively, negative and positive eugenic practices which have now been 'privatised' by industry.[42]

Eugenic movements have played a prominent role in the politics and history of much of the present century. They justified the devastation of indigenous populations by colonising Europeans, and included the introduction of apartheid in South Africa, and the genocide of Jews in Nazi Germany. Hundreds of thousands of U.S. citizens, declared 'feeble-minded', were forcibly sterilised between 1924 and 1974; and the scheme was planned and backed by scientists.[43] Eugenic ideology is responsible for the continuing discrimination against racial minorities and all politically dispossessed groups in the world today. Major concerns about population increase are consistently directed at human populations that are non-white; whereas the real issue is the unequal distribution of resources, of which the well-to-do in predominantly-white developed countries are consuming a disproportionately large share. Finally, one cannot be complacent about the dangers of state-sanctioned eugenic practices even today, as China legislated in 1996 for the compulsory termination of pregnancies diagnosed positive for genetic diseases.

Enclosing the 'Intellectual Commons'[44]

There are yet other repercussions to the current pre-eminence of genetic engineering biotechnology. By defining innovation - for the purpose of financial reward through patenting - as something done within the dominant scientific tradition of Northern Europe, the TRIPS proposal effectively excludes *all* other knowledge systems, especially those in the Third World, but also indigenous or folk wisdom in the North. It also effectively marginalises *any other alternative framework.* Public funding for scientific research since the early 1990s, within most developed countries, has been disproportionately biased in favour of product-oriented genetic engineering biotechnology.

While many areas of basic science are no longer funded, disciplines ranging from embryology and ecology to psychology and anthropology have, one by one, succumbed to the dominant reductionist mindset of genetic determinism. The pluralistic open enquiry that has long been the ideal of science is fast becoming obsolete. Some molecular geneticists have privately told me that they have become increasingly disillusioned within a system that judges excellence on the number of patents owned rather than on the advancement of science.

Genetic engineering biotechnology has effectively reduced the life sciences to a monolithic intellectual wasteland of genetic determinism. It is a "de-intellectualisation of civil society, so that the mind becomes [subjugated to] a corporate monopoly."[45]

This is the real beginning of the brave new world, where ideological control is diffuse, yet automatic and complete, because all effective opposition has been obliterated. That this is all done under the guise of freedom and democracy within 'free' economies in the global 'free-trade' regime of the WTO makes it all the more sinister, and all the more difficult to grasp hold of and to resist. As distinct from the openly totalitarian regimes of the former socialist and communist states, there is no dictator in charge, *there is no one really making decisions*, rational or otherwise. Instead, there are merely automatons driven by a sense of anxiety, isolated individuals driven by a need to amass wealth today and governments driven by the need to remain in power, against the insecurities of the morrow. No one is in charge, especially not those running the big corporations which they do not own, who are accountable to no-one except the shareholders and who care nothing about their companies, except for the share price of their holdings. There are no bosses on whom to lay the blame, only faceless managers who have no real stake in society. In such a system, each person is a piece of floating flotsam adrift at sea, swept on by the tide of events, until the moment of oblivion.

"Monocultures of the Mind"

The failure of the Green revolution is now generally acknowledged. Large-scale monoculture crops and the accompanying use of agrochemicals led to the erosion of indigenous biodiversity, environmental destruction, displacement of indigenous farmers and widespread poverty. Many Third World countries have since devoted major efforts to restoring the environment and to regenerating indigenous biodiversity. They are doing this through a revival of traditional, organic farming, organic methods which are proving to be sustainable and to have a much higher productivity than western monoculture techniques.[46] With hindsight, remembering the example of the Green revolution, why is so much hope pinned on gene biotechnology? It is an even more reductionist ideology producing even more genetically uniform monocultures with accompanying agrochemicals. The significant difference lies in the added danger of genetic pollution and genetic perturbation of ecosystems, which, unlike chemicals, is a self-perpetuating, self-amplifying process that will be impossible to recall.

Sustainable agriculture is also increasingly being practised in northern countries as decades of mechanisation and heavy dependence on agrochemicals have led to declining soil productivity, deteriorating environmental quality, reduced profits and threats to human and animal health. The 1989 report of the National Research Council of the U.S. Academy of Sciences emphasised the development and use of alternative farming systems as a means of increasing productivity and decreasing environmental damage, and estimated that pesticide use could be reduced 75 percent in ten years without loss of productivity.

There are many variants of sustainable agriculture, using a combination of modern and traditional methods, all characterised by a holistic, "system-level approach to understanding the complex interactions within agricultural ecologies."[47] Recent surveys indicate that sustainable agriculture not only overcomes all the problems of conventional, mechanised farming, but is also 22% more profitable. As ecologist Cavalieri concludes, "Sustainable agriculture is an essential goal for a viable future. It's time to put the emphasis on the real means that will get us there."[48] Despite this, the U.S. Department of Agriculture provided less than $5 million to research in sustainable agriculture compared to the $90 million it allocated to gene biotechnology in 1994.

Vandana Shiva, theoretical physicist turned political activist, has been in the forefront of the Third World's struggle (with the Third World Network, TWN) against the exploitative and destructive policies of the North. In the process, she came to realise the pervasive influence of the reductionist ideology - 'monocultures of the mind'[49] - in shaping the policies of the North. In particular, she sees redefining the life sciences as an important part of the struggle. I have come to realise just how much I agree with her.

Notes

1. The Asilomar Declaration was issued as the result of a Conference held in Asilomar, California. It drew attention to the potential risks of recombinant DNA technology. This resulted in the first guidelines issued by the U.S. National Institute of Health (NIH). The city of Cambridge, Massachusetts, debated recombinant DNA research and issued a moratorium pending the outcome of a citizen's review. In 1977, the first law regulating recombinant DNA research was enacted in Cambridge, Mass.

2. See Ho, 1993, 1995a,b, 1997a.

3. Regal, 1994

4. Snow, 1962, p. 138.

5. Hubbard and Wald, 1993.

6. "When the price is wrong", *The Guardian,* Thursday February 27, p. 2, 1997

7. From *Government Response to the Second Annual Report of The Government's Panel on Sustainable Development, Jan. 1996*, March 1996, p.16. Note that this is the Government's response to its own Panel. The reason being that they came up with a Report recommending, among other things, a much more cautious approach to the commercialisation of genetic engineering biotechnology than that taken by the Government.

8. The Human Genome Initiative was established in 1988 by the U.S. National Institute of Health (NIH), under the directorship of James Watson, joint Nobel Laureate for his discovery of the DNA double helix with Francis Crick in 1953.

9. John Moore's spleen was removed as part of the treatment for leukemia. Unknown to him, the physician at UC LA developed and patented a permanent cell line from it, and two companies, Sandoz and Genetics Institute, took out licences. See Burrows, 1996.

10. Shiva, 1994.

11. See Nijar and Chee, 1994.

12. Egziabher, 1994.

13. Cropper, 1994.

14. "Biotech firm 'embarrassed' by leaked plant deal", Rob Edwards, *New Scientist* 29 June, p. 7, 1996.

15. See McNally and Wheale, 1996.

16. "Patent threat to research" correspondence, signed by 9 scientists in the UK *Nature* 384, 672, 1997

17."Patents versus transplants" correspondence, signed by 16 scientists representing the Bone Marrow Transplant Services, Cord Blood Banks and others. *Nature* 382, 108, 1996.

18. See "Letter to Madeleine Albright" initiated by Kristin Dawkins and signed by numerous non-government organisations and individuals from all over the world, protesting that the US has no right to use its commercial power to influence legislative processes in other countries.

19. See for example, Aldridge, 1996, written by a science writer trained in genetic engineering biotechnology.

20. Meister and Mayer, 1994.

21. Palca, 1986; also McNally, 1995.

22. See Belonga, *et al*, 1990, and Mayeno and Gleich, 1994.

23. Nordlee *et al*, 1996.

24. Inose and Murata, 1995.

25. Skogsmyr, 1994.

26. Mikkelsen *et al*, 1996.

27. Holmes and Ingham, 1994.

28. The UNEP Cairo Expert Panel Report, May, 1995,

29. *Biosafety, Scientific Findings and Elements of a Protocol, Report of the Independent Group of Scientific and Legal Experts on Biosafety*, Third World Network, 1996.

30. TWN has played a major role in informing UN delegates on many world issues throughout the 1990s, by providing powerful speakers, running seminars, and preparing briefing papers for circulation at UN conferences. Public interest organisations (ngos) in general are indispensable to the democratic process of the UN. That is because delegates from all countries, both North and South, are typically badly briefed and have little understanding of the issues. That makes it very easy for small groups with vested interests to take over the Conferences. TWN is one of the most influential and knowledgeable public interest organisations working at the UN. They also publish a magazine, *The Third World Resurgence*, which has a world-wide circulation of 100,000.

31. See Chee, 1996.

32.Letter to President Clinton from 40 Associations including biotech companies, food growers and farmers, dated 18 June, 1997. Reported by Reuter June 19. I am grateful to Phil Bereano of the Union of Concerned Scientists of the US and Jaan Surkula of Sweden, for bringing the new item to my attention.

33. See Burrow, 1995.

34. Mellon and Rissler, 1995

35. Miller, 1995.

35. Personal communication, Beth Burrows of the Edmonds Institute, USA.

37. Letter to the Department of the Environment from myself, 14 Feb. 1997.

38. Jaan Suurkula's homepage <http://home1.se/~w-18472/indexeng.htm> contains many informative, up-to-date articles on food safety.

39. This terms is taken from a paper by Hubbard, 1995.

40. See Hubbard and Wald, 1993.

41. *GenEthics News*, issue 3, pp.6-7, 1994.

42. This was predicted by Harvard Marxist geneticist Richard Lewontin in 1985 (see Lewontin, 1985).

43. "Too far, too fast?", Tom Wakefield, *The Guardian*, Wednesday March 5, p.4, 1997.

44. Shiva *et al*, 1997.

45. Shiva, 1994.

46. Perlas, 1994, 1995.

47. Reganold *et al*, 1990.

48. See Cavalieri, 1994.

49. Shiva, 1993.

Chapter 3

The Science that Fails the Reality Test

Reductionist science has failed the reality test because it has been shown not to work in so many cases. Genetic determinism, the most insidious form of reductionist science, also fails the reality test on the basis of scientific evidence. But establishment mainstream scientists have obfuscated and misread the evidence to serve industry and the status quo. The dangers of the mismatch between a powerful set of techniques and an outmoded, discredited ideology guiding its practice should not be underestimated.

Reductionist Science has Failed to Work

In a sense, reductionist science has already failed the reality test simply because it has been shown *not to work* in so many cases. The list includes the Green revolution, eugenics and nuclear energy mentioned in previous chapters. To those we can add the failure of reductionist science to recognise links between mad cow disease and the human neuropathy, CJD; between severe pesticide-poisoning and organophosphates; between chloro-fluro-carbons and the loss of the ozone layer. And many of us still recall the horror of thalidomide, rushed on to the market as a sedative for pregnant women without adequate tests, and eventually withdrawn in 1961 after 8,000 babies had been born with severely truncated limbs.

I do not deny that reductionist science has worked, and worked very well, in solving problems in isolation. What it is notoriously bad at is taking proper account of the organic interconnections, the ecological and social interrelationships, that sustain the living system as a whole.

What I want to show is that genetic determinism, the most insidious form of reductionist science, has failed the reality test also on the basis of scientific findings. So, the science that does not work also happens to be wrong, by the criteria of science itself. This would be entirely obvious were it not the case that there has always been an establishment of mainstream scientists purporting to speak for all scientists, effectively excluding all dissenting views and always appearing to favour the *status quo*, much to the comfort of the powers-that-be. In every one of the failures of science listed above, it was the *other* scientists who provided the evidence as well as the counter-arguments that ultimately proved the orthodox opinion wrong. That is why the blanket condemnations of science and scientists that increasingly appear in the popular media are misdirected. Journalists in the popular media ought to work harder to find scientists with dissenting views and to encourage open debate, rather than promote indiscriminate anti-science sentiments. I shall show how genetic determinism fails the reality test within science itself. But first, let us examine what genetic engineering entails.

The Genetic Engineering Revolution

Genetic engineering is a set of techniques for modifying and recombining genes from different organisms; it is also referred to as recombinant (rDNA) technology. DNA, the genetic material, is a very long chain-like polymer made up of many, many thousands of simpler units joined end to end. The units differ in the organic bases they contain, of which there are four, represented by the letters, A, T, C, and G (Adenine, Thymine, Cytosine and Guanine). The sequence in which the bases occur differs for each DNA molecule, and that accounts for the specificity of the genetic 'message' it encodes. As we shall see in a later chapter, the bases are like the alphabet of a language which can be composed into words, and the words in turn strung into a sentence. Each DNA molecule is packaged into a linear structure, a *chromosome*. Each cell can have one or more chromosomes; for example, a bacterial cell has one chromosome, whereas the human cell has 23 pairs of chromosomes. A *gene* is a stretch of DNA on the chromosome, usually about a thousand units in length, which has a defined function. It might code for one of the thousands of proteins present in our cells, or it might work as a signal for making that protein. (You will see in Chapter 8 why this is a simplistic view of the gene but, for the moment, it lets me get on with the narrative.)

Genetic engineering originated in the 1970s as the result of the development of several techniques. The first, DNA sequencing, allows the sequence of bases in any stretch of DNA to be determined. The second technique is the making of recombinant DNA in the test-tube using enzymes isolated from microorganisms to cut and join pieces of DNA together. This enables geneticists to put foreign genes into *viruses, plasmids,* or *mobile genetic elements,* all of which are pieces of parasitic DNA that can infect cells and multiply in them, or insert themselves into their chromosome and replicate with the host cell. Hence, by cutting and joining bits of viruses, plasmids and mobile genetic elements together, appropriate *vectors* are made for transferring genes from a donor species to a recipient species that does not naturally interbreed with it. The third technique is the chemical synthesis of DNA of any desired base sequence. A fourth technique, the Polymerase Chain Reaction (PCR), discovered in 1988, allows specific gene sequence(s) in a mixture to be rapidly replicated many tens or hundreds of thousand times, and is extensively used in forensic DNA fnger-printing.

I should, right away, dispel the myth that genetic engineering is just like conventional breeding techniques. It is not. Genetic engineering bypasses conventional breeding by using the artificially constructed vectors to multiply copies of genes and, in many cases, to carry and smuggle genes into cells. Once inside cells, these vectors slot themselves into the host genome. In this way, *transgenic* organisms are made carrying the desired *transgenes.* The insertion of foreign genes into the host genome has long been known to have many harmful and fatal effects including cancer;[1] and this is borne out by the low success rate of creating desired transgenic organisms. Typically, a large number of eggs or embryos have to be injected or infected with the vector to obtain a few organisms that successfully express the transgene (see Chapter 11).

The most common vectors used in genetic engineering biotechnology are a chimaeric recombination of natural genetic parasites from different sources, including viruses causing cancers and other diseases in animals and plants, with their pathogenic functions 'crippled'. These are tagged with one or more antibiotic resistance 'marker' genes, so that cells transformed with the vector can be selected. For example, the vector most widely used in plant genetic engineering is derived from a plant tumour-inducing plasmid carried by the soil bacterium *Agrobacterium tumefaciens.* In animals, vectors are constructed from *retroviruses* causing cancers and other diseases. A vector currently used in fish has a framework from the Moloney murine leukaemic virus, which causes leukaemia in mice, but can infect all mammalian cells. It has bits from the Rous Sarcoma virus, causing sarcomas in chickens, and from the vesicular stomatitis virus, causing oral lesions in cattle, horses, pigs and humans.[2] Such mosaic vectors are particularly hazardous. Unlike natural parasitic genetic elements which have various degrees of host specificity, vectors used in genetic engineering, partly by design and partly on account of their mosaic character, have the ability to overcome species barriers, and infect a wide range of species. Another obstacle to genetic engineering is that all organisms and cells have natural defence mechanisms that enable them to destroy or inactivate foreign genes, and transgene instability is a big problem for the industry, as I shall describe in more detail in chapters 8, 9 and 10. Vectors are now increasingly constructed to overcome those mechanisms that maintain the integrity of species. The result is that the artificially constructed vectors are especially good at carrying out horizontal gene transfer.

Let me summarise why genetic engineering differs radically from conventional breeding methods:

1. *Genetic engineering recombines genetic material in the laboratory between species that do not interbreed in nature.*
2. *While conventional breeding methods shuffle different forms (alleles) of the same genes, genetic engineering enables completely new (exotic) genes to be introduced with unpredictable effects on the physiology and biochemistry of the resultant transgenic organism.*
3. *Gene multiplications and a high proportion of gene transfers are mediated by vectors which have the following undesirable characteristics:*

   a. Many are derived from disease-causing viruses, plasmids and mobile genetic elements - parasitic DNA that have the ability to invade cells and insert themselves into the cell's genome, causing genetic damage.

   b. They are designed to break down species barriers so that they can shuttle genes between a wide range of species. Their wide host range means that they can infect many animals and plants, and in the process pick up genes from viruses of all these species to create new pathogens.

   c. They routinely carry marker genes for antibiotic resistance, which is already a big public health problem.

   d. They are increasingly constructed to overcome the recipient species' defence mechanisms that break down or inactivate foreign DNA.

Genetic engineering not only makes it possible for geneticists to manipulate genes, it also happens to be a powerful research tool that enables geneticists to study the genome of organisms in ways that were not possible before. The scientific community was totally unprepared for the plethora of new findings that came to light. They sent shock wave after shock wave through the foundations of classical genetics. Just as quantum physics replaced the classical Newtonian framework at the beginning of the present century, the *new genetics* that came in the wake of genetic engineering overturned every preconceived notion of the old genetic paradigm. Genetic determinism has indeed failed the reality test.

Paradoxical as it may seem, genetic engineering is possible precisely because the paradigm of genetic determinism is invalid. Thus, many of the promises of genetic engineering biotechnology can never be fulfilled because the paradigm is an erroneous, reductionist representation of organic wholeness and complexity. The many problems which have arisen with genetic engineering biotechnology are indicative of that fundamental error in judgment on the part of those who see it as the solution to all the problems that humanity face today. When biotech stocks fell to their doldrums in 1994, *Business Week* reported,

"The industry is still peddling dreams,…." From Wall Street's perspective, "the industry hasn't worked, and the likelihood of success is lower…A long list of products has failed clinical trials, even the remaining handful that got through are not without problems."[3]

This remains true today; "…overall the industry has been so consistently disappointing that laymen should stay away lest they get fleeced….And until the gene-bending gods can separate the hype from the glory, they're not getting any of my savings."[4]

The industry is still peddling dreams: cure for cancer, designer babies, cloning and other means to immortality. It is preying on the very illness and anxieties generated by a society dominated by reductionist science.

I was recently involved in a debate on genetic engineering biotechnology organised by the Society of Chemical Industry at Cambridge University. At dinner afterwards, I found myself sitting next to the CEO (Chief Executive Officer) of a biotech company. In an unguarded moment, he confessed that he personally didn't feel happy about his company's involvement with biotechnology but what could he do? It was the system, and mortgages had to be paid. He was coping, he said, by practising Transcendental Meditation, unlike his colleagues, most of whom were on Prozac.

Financial ruin is perhaps a small price to pay for misjudging genetic engineering biotechnology, when the future of the planet and all its inhabitants is at stake. The dangers of the mismatch between a powerful set of techniques and an outmoded, discredited ideology guiding its practice should not be underestimated. It also constitutes the major stumbling block to a rational debate on genetic engineering biotechnology and its legitimate spheres of application. With that in mind, let us look at the paradigm of genetic determinism and the new genetics in turn.

The Paradigm of Genetic Determinism

A paradigm is a mindset; it is a comprehensive system of thought and practice developed around a key idea or theory. A scientific paradigm is obviously built around scientific theories, but it can be so pervasive as to spill over into all other disciplines, and to permeate the popular culture at large. Genetic determinism is of this nature. It portrays genes as the most fundamental essences of organisms. It supposes that, while the environment can be moulded and reshaped, biological nature in the form of genes is fixed and unchanging and can be sorted from environmental influence. Further, it assumes that the function of each gene can be defined independently of every other. It is on such a basis that the Human Genome Project promises to unravel the genetic programme for making a human being. James Watson, the first Director of the Human Genome Organisation (HUGO), set the tone, "We used to think that our fate was in the stars. Now we know, in large measure, our fate is in our genes."[5]

The twin pillars of genetic determinism are Darwin's theory of evolution by natural selection and the gene theory of heredity as developed by Mendel, Weismann, Johannsen and others. Darwin proposed that evolution occurs by natural selection, in which nature effectively 'selects' the fittest in the same way that artificial selection practised by plant and animal breeders ensures that the best, or most desirable, characters are bred or preserved.[6] The ideology of natural selection is clear: those that do well and survive to reproduce are naturally favoured with superior qualities that can be passed on, like a legacy, to the next generation. In the same way, those with inferior qualities are eliminated. Darwin's theory lacked a mechanism of heredity and variation. This was supplied by Mendel, who proposed that the (Darwinian) qualities inhere in constant factors (later called genes) determining the organisms' characters which are passed on to the next generation during reproduction, and that variations are generated by rare random *mutations* in those genes. (A genetic mutation is a change in the base sequence of the stretch of DNA that constitutes the gene.) The combination of Mendelian genetics and Darwinian theory resulted in the 'neo-Darwinian synthesis'. But, from the beginning, the theories were built on abstraction and ideology, as well as a certain blindness to, and misreading of, scientific evidence. This has recurred again and again, and even outright fraud has been committed in the development of the genetic determinism that we have today, as you will read in later chapters.

The paradigm of genetic determinism today is neo-Darwinism writ large. Thus, everything from IQ to criminality can be explained by invoking a gene or genes responsible, which have been naturally selected for or against. And so they can be hunted for in the genome by geneticists and, having been identified, can be selected for or against by human beings, thereby, we are told, extending our choice as human beings.

There are three basic assumptions to the genetic determinist paradigm:

1. *Genes determine characters in a straightforward, additive (i.e., non-interactive) way.*
2. *Genes and genomes are stable and, except for rare random mutations, are passed on unchanged to the next generation.*
3. *Genes and genomes cannot be changed directly in response to the environment.*

One has to appreciate that those assumptions have been the bread and butter of mainstream biology for at least a hundred years, rather in the way that Newtonian mechanics had been the foundations of physics in the pre-quantum era. *All those assumptions have been contradicted by scientific findings.* Some of us have argued that they were untenable even before the recombinant DNA era. But none of us was prepared for the surprises that rDNA research has turned up within the past 20 years.

Assumption 1 is clearly contradicted by everything that is known about metabolism and genetics for at least 40 years, and no biologist would *admit* to believing in it. However, it is *logically* part and parcel of the genetic determinist science *on which the practice of commercial genetic engineering is based.*

Organisms, including human beings, have tens of thousands of genes in their genome. Each gene exists in multiple variants (you will see exactly how many in Chapter 12). One of the main functions of genes is to code for the thousands of enzymes catalysing thousands of metabolic reactions in the body that provide the energy to do everything that constitutes being alive. These metabolic reactions form an immensely complicated network in which the product of one enzyme is processed by one or more other enzymes. Thus, no enzyme (or gene) ever works in isolation. Consequently, the same gene will have different effects from individual to individual because the other genes in the 'genetic background' are different. So-called single-gene defects - which account for less than 2% of all human diseases[7] - are now proving to be very heterogeneous. Many different mutations of the same gene or of different genes may give the same disease, or not, as the case may be. This has been known for sickle cell anaemia, common in Africans and Afro-Americans and, more recently, for cystic fibrosis, common among Northern Europeans, and a conglomerate of cranio-facial syndromes which includes achondroplastic dwarfism. It has provoked the geneticist reporting in *Nature News and Views* to declare that there is "no such thing as a single gene disease"[8].

Findings which have been steadily accumulating for the past 20 years reveal yet further hitherto undreamed-of complexity and dynamism in cellular and genic processes, many of which destabilise and alter genomes within the lifetime of the organism.[9] This is in direct contrast to the static, linear conception of the 'central dogma' of molecular biology that previously held sway. The central dogma states that the genetic material, DNA, makes RNA (ribonucleic acid) in a faithful copying process called *transcription.* RNA is a nucleic acid like DNA except that the sugar is ribose instead of deoxyribose and, in place of the base thymine (T), it has uracil (U). The RNA then makes a protein by a process of decoding called *translation.* There is strictly a one-way 'information flow' from the genetic message coded in the DNA to RNA to protein, and no reverse information flow is possible (Fig. 3.1a). In other words, proteins cannot determine or alter the transcribed message in RNA, and RNA cannot determine or alter the genetic message in DNA. We shall see that such reverse information flow not only occurs - and in a wide variety of forms - but is a necessary part of how genes function within a metabolic-epigenetic super-network (Fig. 3.1b).

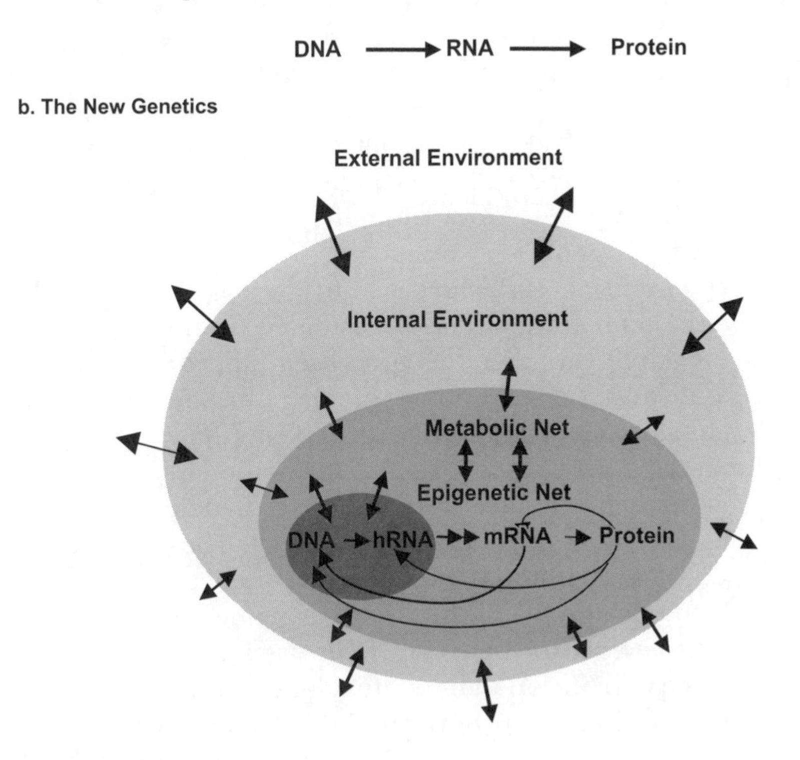

Figure 3.1 Genetics old and new

The Fluid Genome of the New Genetics

A complicated network of feed-forward and feedback processes has to be traversed just to 'express' one gene, i.e. to make a single protein. Genes are found to exist in bits, and the bits must be appropriately joined together to make the messenger-RNA (mRNA). Numerous other proteins take part in making every single protein, chopping and changing, editing and re-coding in a complicated *epigenetic* network which interposes between the genes and the metabolic net, and interlocks with it as an epigenetic-metabolic super-network. Thus, it becomes increasingly difficult to define and delimit a gene, as the super-network ultimately connects the expression of each gene with that of every other.

The genome itself, embedded within the epigenetic-metabolic super-network, is far from stable or insulated from environmental exigencies. A large number of processes appear to be designed especially to destabilise genomes during the lifetime of all organisms, so much so that molecular geneticists have been inspired to coin the descriptive phrase, 'the fluid genome'. Base sequences can mutate, stretches of DNA can be inserted, deleted, or amplified, thousands and tens of thousands of times. The sequences can be rearranged or recombined with other sequences, genes can jump around from one site to another in the genome, and some genes can convert other genes to their own DNA sequences. These processes keep genomes in a constant state of flux in evolutionary time.[10] Genes are found to jump horizontally between species that do not interbreed, being carried by mobile genetic elements or viruses. Ordinary parasites that infect more than one species are also vectors for horizontal gene transfer. A particular genetic element - the *P*-element - has spread to all species of fruit flies in the wild within a span of less than 50 years, probably carried by a parasitic mite.[11] Jumping genes, viruses and vectors for gene transfers are all related genetic parasites. They can help one another jump or mobilise, mutate, exchange parts and infect each other's hosts as a result.

These fluid genome processes are, by no means, entirely stochastic, accidental or meaningless, but are subject to physiological and cellular control, which can be disrupted by gene transfers, such as the insertion of retroviruses that cause cancer. On the other hand, gene jumping, recombination and other alterations of the genome have been found to be also part of the normal physiological response to environmental stress or starvation in non-dividing cells that enables them to cope with new challenges.[12]

Most provocative of all, there is now abundant evidence of (previously forbidden) reverse information flow in the genomes of all higher organisms.[13] Predictable and repeatable genetic changes have been found to occur simultaneously and uniformly in all the cells of the growing meristem in plants exposed to different fertilisers.[14] Similarly, plants exposed to herbicides, insects to insecticides and cultured cells to drugs are all capable of changing their genomes repeatedly by mutations or gene amplifications that render them resistant to the noxious agent.[15]

As a final blow to the genetic paradigm, starving bacteria and yeast cells are now known to respond directly to the presence of (initially) non-metabolisable substrates by mutational changes apparently so specific that they are referred to as 'directed mutations' or, even more provocatively, 'adaptive mutations'.[16] You will have the opportunity to read about the fluid genome in detail in Chapter 8.

Mismatch between Mindset and Reality

There is a serious mismatch between the mindset of genetic engineering biotechnology (with the projected benefits and wealth to be gained therefrom) and the reality of the new genetics. Genetic engineers basically believe that manipulating genes has the potential to solve all the major problems of the world. They reason that, as genes determine the characters of organisms, so, by manipulating the appropriate genes, they can engineer organisms to fulfil all our needs. That can only be true if genes determine the characters of organisms in an uncomplicated way, so that by identifying a gene one can predict a desirable or undesirable trait; by changing the gene, one changes the trait, by transferring the gene, one transfers the corresponding trait, once and for all. The mindset is genetic determinist through and through. Of course, genetic engineers will deny that when confronted. *But that is how the practice is guided, and the basis of the claims that are made.* In other words, genetic engineering biotechnology only makes sense if one believes in genetic determinism. No one would regard it as a good investment if they did not believe in genetic determinism. More to the point, no one would think of it is a good investment if they did not believe that everyone else thought genetic engineering worked in the way it claims.

Let me highlight this mismatch between mindset and reality below, as a convenient summary of the story thus far.

Genetic Engineering Mindset	Reality of Scientific Findings
1. Genes determine characters in linear causal chains; one gene gives one function.	Genes function in a complex network; causation is multidimensional, nonlinear and circular.
2. Genes and genomes are not subject to environmental influence.	Genes and genomes are subject to feedback regulation.
3. Genes and genomes are stable and unchanging.	Genes and genomes are dynamic and fluid, can change directly in response to the environment, and give 'adaptive' mutationsto order.
4. Genes stay where they are put.	Genes can jump horizontally between unrelated speciesand recombine.

The mismatch between mindset and reality is ultimately why genetic engineering biotechnology not only cannot deliver its promises, but also poses such a hazard. I shall refer to a few symptoms here, and leave the detailed diagnosis to subsequent chapters.

Mismatch 1 leads to unrealistic assumptions about the efficacy of gene transfers. Single gene transfers have invariably led to 'unexpected' changes in the recipient organism - 'unexpected' only because reductionists fail to take account of complexity and interconnectedness. Toxins and allergens have arisen as so-called side-effects in transgenic plants and microorganisms, and very sick, monstrous transgenic animals have resulted from having a single gene introduced. This is unsafe for consumers, particularly as regulatory bodies share the same reductionist mindset in risk assessment, so that, often, only information on the genes transferred and the products resulting from those genes is considered necessary and sufficient. It is also morally unacceptable to increase animal suffering, for which there is neither need nor good scientific justification.

Mismatch 2 leads to unrealistic neglect of physiological and environmental feedback regulation. Calgene's Flavr Savr tomato, genetically engineered to improve shelf-life, and the very first *live* transgenic food to be introduced to our supermarkets, has now been withdrawn. Apparently, because it was developed in California, it does not grow properly in Florida.[17] Similarly, Monsanto's *Bt*-cotton crop - engineered with a gene from a soil bacterium producing a toxin against insect pests - did not work properly when it was first planted commercially in Texas in 1996, because it was "too hot". Nor did it work properly in Australia, probably because it was "too cold".[18] The instability of transgene expression due to *gene silencing* and, in some cases, loss of the transgene, is a big problem for the biotech industry, although they are doing their best not to communicate such failures to the public. In April 1997, Monsanto pulled two varieties of genetically engineered canola seeds from the Canadian market after testing revealed that at least one of the patented herbicide-tolerant transgenic varieties contained an 'unexpected' gene. This was after 60,000 bags of the seeds had already been sold throughout Western Canada.[19] Once again, it is the reductionist disregard for complexity and wholeness that is failing. It is unsafe and insupportable as an investment in the long-term. The industry has, in fact, produced little else than dreams. It is still living off the gullible in stock market shares.

Mismatch 3 is particularly relevant to a large class of current transgenic plants with built-in biopesticide, the *Bt*-toxin, Insects were found to develop resistance rapidly when exposed to the toxin. That was the other problem with the *Bt*-cotton crop in Texas. Classical neo-Darwinian theory puts this down to natural selection of pre-existing, rare 'random' mutations. However, the real story is that all cells and organisms have the physiological capability to develop resistance by a wide variety of genetic mechanisms, from multiplying particular genes thousands of times to generating new genes by recombinations or mutations. They can also acquire the genes needed from their friends, which include organisms from other species.

Genes and genomes are inherently fluid and dynamic. It is the failure of reductionist science to recognise that genetic stability is a property, not of the gene transferred, but of the ecological whole in which the organism is entangled (see Chapter 14).

Mismatch 4 is perhaps the most serious of all. It is the failure to recognise that genes do not remain static in the genome once and for all. They can also jump between species that do not normally interbreed, particularly when the conditions are favourable. The instability of the genome is, to a large extent, due to genetic parasites, including mobile genetic elements, plasmids and viruses, that can hop in and out of genomes, replicate themselves and infect other cells. As I have already pointed out in Chapter 1, genetic engineering is inherently hazardous because it uses vectors constructed out of these genetic parasites to facilitate horizontal gene transfer. And horizontal gene transfer is already known to be responsible both for the emergence of new and old pathogens and of multiple antibiotic resistance.

Failures of Reductionist Medicine

If we need more evidence that reductionist science has failed the reality test, it can be found in the area of public health. As is evident from the WHO 1996 report, diseases such as tuberculosis, malaria, cholera and yellow fever are still a major cause of death in many parts of the world, and these diseases are returning to regions where they were on the decline. New diseases also continue to emerge at unprecedented rates from social conditions and environmental disturbances that enable pathogens to gain access to new host populations, or to become more virulent in immunologically-weakened human hosts, suffering from poverty and malnutrition. Many of the agents of infectious diseases rapidly develop resistance to the drugs and chemicals, while new variants continue to arise that escape the protection of vaccines, as we would expect from what we know of the fluid genome. The conclusion of The Harvard Working Group on New and Resurgent Diseases is that,

"Disease cannot be understood in isolation from the social, ecological, epidemiological and evolutionary context in which it emerges and spreads. Indeed, if one lesson has emerged from the spectacular failure of Western medicine to 'eradicate' certain diseases, it is that diseases cannot be reduced to a single cause nor explained within the prevailing linear scientific method: complexity is their hallmark."[20]

Genetic engineering biotechnology greatly exacerbates the problems of new and resurgent diseases in facilitating horizontal gene transfer. It is also diverting attention and resources away from the real causes of ill health, which are poverty with its consequent malnutrition and lack of sanitation, proper housing and clean water supply.

Based on the same reductionist ideology, chemical and pharmaceutical industries are producing ever more exotic chemicals, drugs and cures that have already done more harm in environmental pollution and in physiological side-effects than the conditions they are supposed to overcome or treat. The history of the pharmaceutical industry is littered with failures of wonder-drugs. Iatrogenic diseases, or diseases caused by prescription drugs, are estimated to result in 2 million Americans being hospitalised each year, and 180,000 deaths.[21] Do we need even more of the same? Is it not time to stop profiting from the ill health that has already been created by the over-use and abuse of drugs?

Implications of the New for Heredity

The genetic determinist paradigm has collapsed under the weight of its own momentum in the findings of the new genetics. The genes are far from being the constant essences of organisms, whose effects can be neatly separated from one another or from the environment. There is, furthermore, no constant genetic programme or blueprint for making the organism, for the genome itself can also change in the course of development. For an organism, being alive depends on a dynamic balance of feedback interrelationships between it and its environment which extends through its entire physiological system to its genes. Heredity - the stability of reproducing the same life cycle - is the property not so much of the genes as of the whole system of the organism within its ecological environment. All of its parts are, ideally, maximally responsive and communicative with every other part,[22] and must change as appropriate in the maintenance of the whole. Heredity, therefore, does not reside in our genes any more than it resides in the environment and cultural traditions that we have created, and which are also passed on to future generations. Heredity is the whole way of living, characteristic of the species. The biological and socio-ecological are mutually, inextricably entangled. Consequently, our fate is written neither in the stars nor in our genes, for we are *active participants* in the evolutionary drama.[23] The new genetics that underpins genetic engineering biotechnology belongs with a paradigm of organic wholeness and complexity emerging in many areas of contemporary research in the west, which is reaffirming the universal wisdom of traditional indigenous cultures all over the world.[24] Hence, we can choose to shape the socio-ecological conditions for our biological health and wellbeing, rather than persist in futile and hazardous attempts to manipulate our genes and those of other species.

The Struggle to Reclaim Holistic Ways of Life

The collapse of the genetic determinist paradigm is both symptomatic and symbolic of the collapse of the reductionist world view. It is significant that, in opposition to the patenting of life, French geneticist Daniel Cohen, a prominent figure in the Human Genome Project, made the first move to offer a wide range of DNA sequence data obtained in his laboratory to the United Nations as the property of humanity to use freely for any appropriate purpose.[25]

The debate over genetic engineering biotechnology is not about disembodied, objective, ivory-tower scientific knowledge. Knowledge is what people everywhere else in the world live by. The western ideal of being objective is misplaced, for it implies that one must be a completely detached, unfeeling observer outside nature. Within the participatory framework of all other knowledge systems, the ideal of objectivity in knowledge is to be maximally communicative and connected within the nature that is the object of our knowledge, which we, as both knower and actor, participate in shaping.[26] The present opposition to genetic engineering biotechnology is thus also a concerted struggle to reclaim holistic world views and holistic ways of life that are spontaneous, pluralistic, joyful, integrative, constructive and life-sustaining.

Notes

1. Wahl, *et al*, 1984; see also relevant entries in Kendrew, 1995.

2. Lin *et al*, 1994

3. Hamilton and Carey, 1994.

4. "Bearish on biotech" Daniel Kadlec *Time* March 10, p.52, 1997.

5. Quoted in *GeneWatch* 9, p.5, Nov. 1994.

6. Darwin, 1859.

7. See Strohman, 1994, for an excellent critique of reductionist, linear concepts in health and disease.

8. Mulvihill, 1995.

9. The evidence is extensively reviewed by a number of authors beginning more than ten years ago; see Steele, 1979; Dover and Flavell, 1982; Pollard, 1984; Ho, 1987; Rennie, 1993; Jabonka and Lamb, 1995. See also Chapter 8.

10. See Dover and Flavell, 1982.

11. Rennie, 1993.

12. Foster, 1992.

13. Rothenfluh and Steele 1993.

14. Cullis, 1988.

15. Pollard, 1988.

16. See Foster, 1992; Symonds, 1994.

17. Information sheet, Pure Foods Campaign.

18. "Bt cotton fiascos in the U.S. and Australia", Biotechnology Working Group Briefing Paper No. 2, BSWG, Montreal, Canada, May 1997.

19. See *Manitoba Co-Operator*, 24 April, 1997; also *The Ram's Horn*, No. 147, April, 1997.

20. The Harvard Working Group on New and Resurgent Diseases (1995). New and resurgent diseases. The failure of attempted eradication. *The Ecologist* 25, 21-26.

21. Natural Law Party briefing paper; see also Brennan *et al* , 1991.

22. This ideal is satisfied when the system is *coherent*. For the biophysics of coherence, see Ho, 1993: 1995a,b; 1997a

23. See Ho, 1988a.

24. Ho, 1993.

25. "What's wrong with neo-Darwinism?" Brian Goodwin, *THES* May 19, p.18, 1995.

26. See Ho, 1993 and 1996b for a rigorous argument that participatory knowledge is the only rational knowledge, according to contemporary western science.

Chapter 4

The Origins of Genetic Determinism

Genetic determinism has a very strong hold over the public imagination. Its ideological roots reach back, deep within the collective unconscious of our culture, to Darwin's theory of evolution by natural selection, which is itself a product of the socio-economic and political climate of Victorian England. The ruling classes believed in progress through competition in the 'free market' created by the military might of the imperialist army, while positivistic philosophy preached the triumph of mechanical materialism over religion and other romantic ideas that there might be any 'purpose' in life. The belief in the constancy and fixity of genes substituted for the belief in an immortal soul when science replaced religion.

Deconstructing the Old

When I left molecular genetics ten years ago, I felt I was entering a wonderfully constructive period in science when we would articulate a holistic world view *based on contemporary western science.* I have since come to realise that our efforts at deconstructing the old paradigm were not good enough. Most of all, I now know how much it matters to the issues raised by genetic engineering biotechnology and the way they will shape our lives. And how important it is that we really exorcise the ghost of the old paradigm, to set ourselves free, before we can wholeheartedly accept the new.

Genetic determinism has a very strong hold over the public imagination. That is because its ideological roots reach back, deep within the collective unconscious of our culture, to Darwin's theory of evolution by natural selection, and beyond; for Darwin's theory is itself a product of the socio-economic and political climate of Victorian England. Victorian England saw the rise of capitalism, the expansion of trade by imperial conquests. Its ruling classes, in particular, believed in progress through competition in the free market or, more accurately, through the 'free market' created by the military might of the imperialist army. It was also the era of positivistic philosophy that believed in the triumph of mechanical materialism over both religion and other romantic ideas that there might be any 'purpose' in life.

The scholar and historian, Jacques Barzun, maintained that one of the great turning points in modern history came in 1859 when Charles Darwin's *Origin of Species*, Karl Marx's *Critique of Political Economy*, and Richard Wagner's *Tristan and Isolde* made their first appearance. Together, they dominated the epoch, and their theories epitomise a century of thought which continues to the present day.

"Emerging out of an era of Romanticism and flowering in an age of 'scientific thought', their 'mechanical materialism' expresses the prevalent conception of matter as the source and substance of the universe. Feeling, beauty, and moral values are mere illusions in a world of fact, and the human will is powerless against the ineluctable laws of nature and society."[1]

While it is true that mechanical materialism breeds alienation and a sense of powerlessness, particularly in those who do not possess the knowledge, its explicit aim is otherwise. For it firmly believes in the power of abstraction and reduction to make sense of the untidy, intangible complexity of real processes, so that man might better control and dominate nature. Let us begin with the theory of evolution.

Lamarck, Darwin and Evolutionary Theory

Evolution refers to the natural (as opposed to supernatural) origin and transformation of the living inhabitants of the planet Earth throughout its geological history to the present day. Many in the west have speculated on evolution since the time of the Greeks. The ideas which have come down to us, however, originate in the European Enlightenment. That period saw the beginning of Newton's laws of mechanics, mathematics and other modern western scientific ideas, including John Ray's concept of species and Carl Linnaeus' system for classifying and naming organisms, which is still adopted today.

Most of all, the power of rational thought (in science) to explain the material universe presented a deep challenge to received wisdom, especially the biblical account of creation according to the Christian Church. Evolution by natural processes - as opposed to special creation by God - was already in the minds of most educated people. Linnaeus came to accept a limited transformation of species later in his life; other prominent figures who wrote on the possibility of evolution include the naturalist, G.L. Buffon and Charles Darwin's grandfather, Erasmus Darwin.

The first comprehensive theory of evolution is due to Jean Baptiste de Lamarck, who was very much a product of the Enlightenment, both in his determination to offer a naturalistic explanation of evolution and in his systems approach. Thus, he dealt at length with physics, chemistry and geology before embarking on presenting evidence that biological evolution had occurred. He also suggested a mechanism of evolution, whereby new species could arise through changes in the relationship between the organism and its environment in the pursuance of its basic needs, thereby producing new modifications in its characteristics that become inherited after many successive generations.[2]

Lamarck's theory was widely misrepresented as merely the inheritance of acquired characters, or caricatured as changes resulting from the wish fulfilment of the organism. Half a century later, Charles Darwin was to include a number of Lamarck's ideas in his own theory of evolution by natural selection, but without due attribution. The theories of evolution and heredity are closely intertwined in their historical development. Just as evolutionists needed a theory of heredity, so plant breeders in the eighteenth century who inspired Mendel's discovery of genetics were motivated by the question of whether new species could evolve from existing ones (see Chapter 5).

In accounting for change or transformation, it is also necessary to locate where constancy or stability resides. Charles Darwin's theory of evolution by natural selection says that, given that organisms can reproduce more of their numbers than the environment can support, and that there are variations that are inherited, then, within a population, individuals with the more favourable variations will survive to reproduce their kind at the expense of those with less favourable variations. The ensuing competition and 'struggle for life' results in the 'survival of the fittest', so that the species becomes better adapted to its environment. And if the environment itself changes in time, there will be a gradual, but definite change in the species.[3] Thus, nature effectively 'selects' the fittest in the same way that artificial selection practised by plant and animal breeders ensures that the best, or the most desirable characters are bred or preserved, while those that are undesirable or 'unfit' are eliminated.

In both cases, new varieties are created after some generations.

It is to Darwin's credit that he realised natural selection could not explain everything, so, *in addition* to natural selection, he invoked the effects of use and disuse, and the inheritance of acquired characters in the transmutation of species - both previously proposed by Lamarck. The effects of use and disuse simply mean that if the organism makes use of any part of its body habitually, that part will develop and work better; conversely, any part that is under-used will atrophy or shrivel away. For example, people who train for the marathon will have strong muscles in their legs that do not fatigue easily.[4] Conversely, astronauts cannot tolerate low gravity conditions for very long before their muscles begin to suffer degeneration from disuse. The effects of use and disuse are now well documented. But whether those effects are inherited is not yet known. It is clear, however, that those Lamarckian ideas do not fit into the theory of natural selection, and are overwhelmingly rejected by Darwin's followers, who regard the lack of a theory of heredity and variation as the weakest link in the argument for natural selection.

It was Mendel who provided the missing link in Darwin's argument. Mendel published his work in the same year that Darwin's *Origin of Species* appeared. Yet, in direct contrast to the instant fame that Darwin's publication brought to its author, Mendel's theory, which conforms even more to the mechanical materialism model (as will be described in detail in the next chapter), languished in the archives for more than 40 years before it was rediscovered at the turn of the present century. Darwin himself was sent a copy of Mendel's publication, which remained in his library, uncut and unread. This only goes to show that genetics by itself would not excite anyone's passion. Rather, it is the presumed value (referred to as 'fitness') of the characters determined by the genes that is crucial in accounting for and justifying the existing social order. Winners and losers alike in the competitive 'struggle for life' are what they are by the ineluctable necessity of natural law.

Genes, as unchanging essences of the organism, also take on the symbolic significance of the soul that was lost when science replaced fundamentalist religion. By the time Mendel's work was rediscovered, the German biologist August Weismann had already identified the material basis of heredity as the 'germplasm' or germ cells which became separate from the rest of the animal's body in the course of early development. Weismann was inspired by the same zeal for mechanical materialism. His explicit aim was to reduce heredity to its material, mechanical causes. He was studying the development of insects, the embryos of which form separate germ cells in the posterior pole very early in development. This gave Weismann the idea that hereditary influence - later to be equated with Mendelian genes - could be passed on unchanged from one generation to the next through the germ cells. Weismannism and its implied 'immortality of the germplasm' symbolises the reassuring persistence of the eternal soul, of order and fixity in the face of change; and has had a very strong and lasting influence in western thought.

After Mendel's theory was rediscovered simultaneously by several scientists, including its major champion, the British biologist, William Bateson, a fierce debate ensued. This revolved around the question of whether evolution was continuous and gradual, with natural selection acting on small insensibly graded variations, as Darwin had proposed, or it occurred by large variations caused by mutations, which were then subject to natural selection. On the whole, the Mendelians believed in the efficacy of large, discontinuous variations arising from mutations, or *sports*, whereas their opponents, the *biometricians* believed that continuous variations were the stuff of evolution. We shall see how the debate was resolved in the neo-Darwinian synthesis in Chapter 6.

The Epigenetic versus the Genetic Paradigm

History has the habit of creating heroes and anti-heroes, and so Darwin triumphed while Lamarck bore the brunt of ridicule and obscurity. The reason is that the theories of the two men are logically diametrically opposed. Darwin's theory is natural selection, and selection entails a separation of the organism from its environment. The organism is thus conceptually closed off from its experience, leading logically to Weismann's *barrier* (see below) and the central dogma of molecular biology (Chapter 7) which is reductionist both in intent and actuality. The life experiences of the organism and the organism itself are both negated, as only its genes are of any consequence, both in development and in evolution. This fatalism is inherent to the genetic determinist paradigm. Lamarck's theory, on the other hand, is of transformation arising from the organism's own activities and experience of its environment during *epigenesis* or development. Lamarck's theory requires a conception of the organism as an active, autonomous being, which is open to the environment. Its openness is inherently subversive of the *status quo*. No wonder it was suppressed. It also invites us to examine the dynamics of transformation, and the mechanisms whereby the transformation could become 'internalised' in the course of development and evolution. It is consistent with the epigenetic approach[5] that has emerged as an alternative to neo-Darwinism since the late 1970s, to which Ihave contributed and which is fully vindicated by the new genetics. The epigenetic approach is one that takes the organism's experience of the environment during development as central to the evolution of organism. It is potentially always subversive of the *status quo*, which is why it is invariably vehemently denied by the present orthodoxy.

Where did Darwin's Theory come from?

According to usual accounts, Darwin's voyage around the world as a naturalist on board the *Beagle* enabled him to observe the evidence for evolution, but the idea of natural selection as a mechanism to account for evolution came from three other sources:[6]

1. The theological 'argument from design' as popularised by William Paley, which essentially says living things are so perfectly adapted for their way of life that they must have been designed by a supremely intelligent being, or God.
2. Artificial selection, as practised by plant and animal breeders.
3. Thomas Malthus' famous Essay on the Principle of Populations, first published in 1798. This was the same essay that had inspired Alfred Wallace, another naturalist, who is given joint credit with Charles Darwin for having independently proposed the theory of natural selection. Actually, many people had proposed the same idea before, including Social Darwinist, Herbert Spencer, who first used the phrase, "survival of the fittest", and who also got the idea of natural selection from reading Malthus' essay.

Malthus and 19th Century Britain

So, what was Malthus' essay about? It declared that populations always tended to increase in 'geometric ratio', i.e. 2 gives rise to 4, then 8, then 16 and so on, whereas the means of subsistence only seems to increase in 'arithmetic ratio', i.e. 2, 4, 6, 8, and so on, by bringing new land into cultivation. From this, he concluded that mankind must always be subject to famine, poverty, disease and war, unless some means of limiting population could be found. Malthus was reacting against the optimistic ideals of the Enlightenment The French philosopher Condorcet, for example, had believed that through suitable education and exercise of their rational faculties, human beings might be able to ascend to a kind of Utopia, and he hoped that the French Revolution might be the instrument whereby that could be achieved. Malthus argued that it was impossible to realise such ideals because they took no account of the law of populations.

Malthus' essay generated a great deal of controversy. Historian of science and political theory, Phil Regal, proposes that Malthus' essay was meant to placate the masses in Britain, in case they had ideas of revolting against the monarchy like the French.[7] In any case, Malthus was responding to a particular political controversy in England in the late eighteenth century concerning the restructuring of the Poor Laws. There was, at the time, a limited number of workhouses for the aged and the unemployable. For the most part, benefit was paid in the form of 'outdoor relief', i.e. outside the workhouses. Relief was provided by the parishes in the form of doles, family allowances and 'aid-in-wages'. This became known as the Speenhamland system, after the Justice of the Peace of Speenhamland in Berkshire who initiated it. The system was adopted in several other counties, and William Pitt proposed a bill in Parliament that would allow the system to be introduced generally. The purpose was to raise the birth rate in Britain so as to help fight Napoleon![8]

Malthus' essay was also a statement of his objection to the Pitt proposal. He argued that providing extra financial support would merely encourage the poor to breed more, leading to the need for more relief, and everyone would be dragged down together. This convinced Pitt to withdraw the proposal. Despite some liberal attempts to defend Malthus, it is clear that he was speaking on behalf of the ruling classes, using science to bolster his arguments, which then got incorporated into Darwin's theory. Some will, no doubt, recognise Malthus' arguments in the current concerns about population control expressed by Northern countries and directed against the predominantly non-white populations of the Third World. Whereas, the real issues are the continuing exploitation of the poor by the rich, of the South by the North, with their unequal terms of trade and the resultant inequitable distribution of the world's income, of which the overwhelmingly disproportionate larger share is consumed by the well-to-do in the industrialised North.

The ideas of Malthus had a profound effect on social policies in Victorian Britain, which continued, to a large extent, under the Conservative Government which was defeated in the general election of 1997 after 18 years in power. In 1834, the year of Malthus' death, the New Poor Law was introduced, in which 'outdoor relief' was replaced by workhouse relief. Conditions in the workhouses were deliberately made harsh to dissuade people from coming into them. Once in them, the sexes were strictly segregated in order to discourage the production of children.

Malthus' theory of population was not based on any rigorous scientific observations. It arose from ideology bolstered by a dubious mathematical argument. Natural populations are seldom controlled by disease or famine, unless they have been drastically disturbed. That is because social and other environmental factors tend to stabilise natural populations in a balanced ecological community. In Britain itself, it was the Industrial Revolution coupled with aggressive capitalism and the usurpation of common land by the aristocracy that created the poor (surplus) masses. In the same way, many indigenous populations in the Third World had stable, sustainable relationships with their ecological environment before they were displaced by invading Europeans. The glorious British Empire in Victorian times was built on the exploitation of its own masses as much as that of the colonised peoples of the Third World. Darwin and the other major proponents of natural selection were all inspired by a social theory which has no substance other than that it was the ideology of the dominant ruling classes in Victorian England.

Darwinism and Politics of the Present Century

Although Darwin documented at length the existence of variations in animals and plants in nature as well as under domestication, he had little to say on how those variations originated, except that they were 'accidental'. In fact, it was the natural selection of *accidental* variations that made the whole process mechanical and automatic, as Barzun points out. It was a denial of Divine providence and purpose as much as individual human consciousness and will. And yet, the language in which it is couched betrays a schizophrenia: for it glorifies competition and the struggle for life with the 'fittest' ultimately winning the race. The full title of Darwin's book is even more explicit. It proposed *The Origin of Species by Means of Natural Selection, or the Preservation of Favoured Races in the Struggle for Life.* Thus, an impersonal *validating* purpose has replaced God. It is simply the lottery of life which gives you your genetic endowment. It is a theory that inspires a heroic triumphalism in the ruling classes and, at the same time, inculcates in the lower classes a fatalistic acceptance of the existing social order. If you are a rich capitalist who has succeeded in making a lot of money, then it is because you are among the favoured ones in the struggle for life. Similarly, if you happen to be a poor, downtrodden, exploited worker, why then, that's your lot in life, because you were genetically poorly endowed.

Darwinism epitomises the development of a Zeitgeist in 19th century Britain, which in turn lends credibility to some of the most pernicious political ideologies in the present century. As Barzun remarks, " 'Matter' and 'force', particularly when applied to human beings, found some dangerous, simple applications... And when the idea of force is embodied in the notions of Struggle and Survival of the Fittest, it should be expected that men will use these revelations of science as justifications for their own acts."[9] This completes a positive feedback loop between the dominant socio-cultural ideology of 19th century Britain and a scientific theory to which it gave birth. It led, via the Social Darwinists and theories of racial inequality, to the devastation of indigenous populations by invading Europeans in the Third World, to the slave-trade, the segregation in the Southern States of the U.S., apartheid in South Africa, and the persecution of the Jews in Nazi Germany. It lurks behind the continuing discrimination against racial minorities and against all politically dispossessed groups in the world today.

Notes

1. Barzun, 1958, back book cover.

2. Lamarck, 1809.

3. Darwin, 1859.

4. See Ho, 1995a for more details on muscle energetics.

5. See Ho and Saunders, 1979; Ho, 1984a for a *consistently* epigenetic approach as outlined here. Although other approaches (such as Goodwin, 1984, 1994; Strohman, 1993, 1997) place much emphasis on development, they do not accept the actual feedback interrelationship between organism and environment as being crucial to the development and evolution of the organism itself.

6. See Young, 1985.

7. Phil Regal, personal communication.

8. See Oldroyd, 1980, for an excellent account for the social context from which Darwinism arose and its subsequent impacts.

9. Barzun, 1958, p. 92.

Chapter 5

The Birth of Genes

Mendel abstracted away the untidy complexities of living, experiencing organisms for ideal, eternal entities that behave in a simple, logical way. There is an unfortunate tendency to mistake the abstraction for reality - a case of the fallacy of misplaced concreteness - which is where reductionism begins. People start to forget the organism altogether by thinking of it as a mere collection of genes. They thus discount not only how genes can give rise to organisms (if indeed they can), but also how they interact with one another and are absolutely dependent on cellular and ecological contexts.

The Gene Theory of Heredity

The gene theory of heredity attempts to explain how the characters of organisms are inherited in subsequent generations. Thus, it explains not only why organisms which are all alike should 'breed true' when they reproduce among themselves, but also why, when organisms which differ in some characters are interbred, their offspring may be intermediate between the parents or may resemble one or the other parent. Furthermore, all the possibilities may be found in subsequent generations. For example, within families, children may resemble one or other of their parents or neither of them, but resemble a grandparent or an aunt or a distant cousin instead.

It turns out that, in order to account for all those possibilities in sexually reproducing organisms (as non-sexually reproducing organisms merely reproduce identical copies, or *clones*, of themselves) a *particulate* theory of inheritance is required. In other words, there must be factors that can remain separate and discrete, as opposed to those that mix and blend, and lose their identity. It will help to summarise the elements of the theory in its fully-developed form, that is, in the form in which it came to be accepted for the best part of the present century, up to at least the early 1970s. I shall point out later why it does not explain heredity except in a very trivial and limited sense.

Elements of the gene theory of heredity:

1. *The characters of an organism are determined by stable unit factors called genes, each represented in two copies (alleles) which may be the same or different from each other.*
2. *Each organism carries a large number of genes.*
3. *The genes are passed on unchanged from parent to offspring via the germ cells.*
4. *Each germ cell contains a single copy of each gene, so that the precise combinations of genes will vary at random from one germ cell to the next.*
5. *When the germ cells unite at fertilisation, the resulting zygote (fertilised egg) will again have two alleles of every gene.*
6. *The separation of alleles and recombination between alleles of different genes during reproduction account for the resemblances and differences between successive generations.*

It is entirely a matter of chance which combinations of genes are passed on from each parent, and that in turn will determine the characters of each offspring. If we concentrate on the father in a family, he would have had an equal contribution from each of his parents: one allele of every gene from his mother,[1] and the other allele from his father. However, the sperms he produces may contain any combination of alleles: all from his father (paternal contribution), all from his mother (maternal contribution), and every possibility in between the two extremes. So, just by chance, a child that he produces may have half of its genetic make-up from one grandparent, and that may account for why it ends up resembling that grandparent a great deal.

But, both parents make equal contributions to their children, so why aren't the children intermediate in every respect between the two parents? For example, among families where one of the parents is fair-haired and the other dark-haired, it may turn out that none of the children has intermediate hair colour. Instead, the children are either all dark-haired, or half of them are fair-haired, and the other half dark-haired. In order to explain this, additional concepts are required, as follows:

7. *The genetic constitution of an organism is its genotype, which is to be distinguished from its expressed characters, the phenotype.*
8. *The action of one allele may be dominant or recessive to that of*

another, according to whether it is expressed, or not, in combination with the other allele.

Historically, the concepts of dominance and recessivity preceded those of genotype and phenotype. Suppose we represent the dark-haired allele by *D*, and the fair-haired allele by *d*, we can explain the inheritance of hair colour by assuming that *D*, for dark hair is dominant to *d*, for fair hair. What is the phenotype of individuals with the following genotypes: *dd*, *DD* and *Dd*? The answer is, fair (*dd*), dark (*DD*) and dark (*Dd*). The individuals carrying two identical alleles (*dd* and *DD*) are *homozygotes*, or homozygous for the gene, whereas those carrying different alleles are *heterozygotes*, or heterozygous for the gene.

Let us see what happens in a family where both parents are fair-haired. Their genotypes are therefore both *dd*. We can represent the process of reproduction by a diagram as follows:

Genotype of parents:	dd dd
Phenotype:	fair fair
Genotype of gametes:	d d
Genotype of children:	dd
Phenotype:	fair

This neatly explains why the offspring will always be fair-haired, like the parents. The same would apply to the dark-haired homozygotes, *DD*. Homozygotes, therefore, always 'breed true'. They form so-called 'pure lines'.

(Concepts such as 'breeding true' and 'pure lines' are easily pressed into the service of calls for the preservation of 'racial purity' and other racist, eugenicist slogans. They are based on the mistaken assumption that pure lines actually exist. All human populations are genetically diverse, with several common alleles in most genes. For most sexually reproducing organisms, it is impossible to obtain pure lines which, by definition, would have to be homozygous in all their genes. When laboratory experiments are carried out to try to make lines that are homozygous in as many genes as possible by inbreeding, i.e. mating genetically-related individuals, such as siblings with one another or parents with offspring, they tend to die out rapidly from adverse effects collectively referred to as 'inbreeding depression'.)

What happens in families where both parents are heterozygous? Again, let us represent the process of reproduction diagrammatically,

Genotype of parents:	Dd	Dd		
Phenotype:	dark	dark		
Genotype of gametes:	D	d	D	d
Genotype of children:	DD	Dd	dD	dd
Phenotype:	dark	dark	dark	fair

Each heterozygous parent produces two types of gametes, carrying the allele *D* and *d* respectively, with equal probability. Random combination of gametes from the two parents gives equal probability of all possible combinations of genotypes: 1*DD*:2*Dd*:1*dd*. The result is the famous 3:1 Mendelian ratio of dark-haired to fair-haired phenotypes in the offspring. This is explained by saying that the dark allele *D* is *dominant* over the fair allele, *d*. Or, alternatively, the fair allele is *recessive* to the dark allele. This 3:1 ratio holds only on average, and will be approximated more and more closely when a large number of children are produced, or when the children from a large number of similar families are counted. It is quite likely that some heterozygous matings will produce mostly fair-haired or dark-haired children just by chance. I leave it as an exercise for the readers to work out the possibilities in families where one parent is fair-haired and the other dark-haired. (Hint: there are two kinds of such families.)

What happens if neither of the alleles for hair colour is dominant or recessive? In that case, they are referred to as *co-dominant*, the phenotype of the heterozygote will be intermediate, and the ratio of phenotypes will then be exactly the same as that of the genotypes: 1 dark (*DD*): 2 brown (*Dd*): 1 fair (*dd*).

You can already see that the relationship between genes and characters is not completely straightforward even in Mendelian genetics, as some genes can be dominant in expression to others.

In order to illustrate how the alleles are reshuffled at reproduction, let us bring in another gene, an allele of which is responsible for cystic fibrosis, by far the most common genetic disease among Northern Europeans. It is estimated that approximately one in ten in the population is heterozygous, or a carrier of the disease who shows no adverse symptoms. That means the allele giving rise to the disease is recessive to the majority normal, or wild-type, allele. It is conventional among geneticists to represent the recessive allele by a small letter, and the dominant allele by the corresponding capital letter. So, let's call the cystic fibrosis allele *c*, and the wild type *C*. Let us assume that both the parents are heterozygous for the hair colour alleles as well as the cystic fibrosis alleles, and try to predict the result of their reproduction.

Genotypes of parents:	CcDd	CcDd
Phenotypes:	dark, normal	dark, normal
Genotypes of gametes:	CD Cd	cD cd

In order to predict the genotypes and phenotypes of the children, let us use a Punnett square (named after the geneticist who first used it). The possible gametes of one parent are represented in a row at the top, those of the other in a column on the left. All the possible combinations are then filled into the appropriate grid referenced by the column and the row, as follows:

	CD	Cd	cD	cd
CD	CCDD	CCDd	CcDD	CcDd
Cd	CCDd	CCdd	CcDd	Ccdd
cD	CcDD	CcDd	ccDD	ccDd
cd	CcDd	Ccdd	ccDd	ccdd

The ratios of the possible genotypes are:

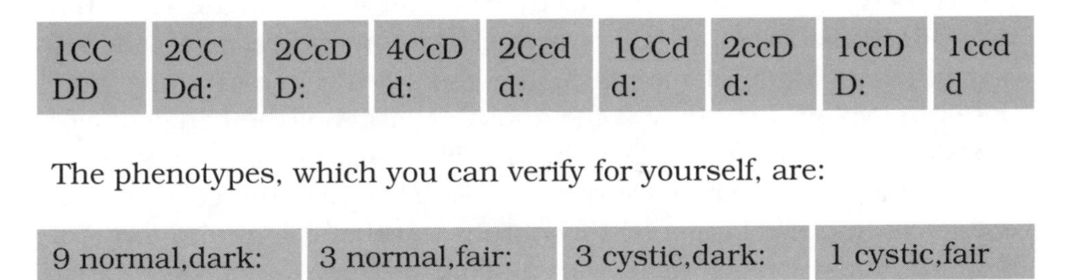

1CC DD	2CC Dd:	2CcD D:	4CcD d:	2Ccd d:	1CCd d:	2ccD d:	1ccD D:	1ccd d

The phenotypes, which you can verify for yourself, are:

9 normal,dark:	3 normal,fair:	3 cystic,dark:	1 cystic,fair

You have now completed the basics of Mendelian genetics, named after its founder, Gregor Mendel, and learned much of the language besides. It is usually presented as the two 'laws': the law of segregation, referring to the separation between the alleles in the formation of gametes, and the law of independent assortment, referring to the random way in which the alleles of different genes are allotted to the germ cells.

Actually the second law is not accurate, as the genes are strung together in groups called linkage groups, later found to correspond to *chromosomes*. Humans have 2 sets of chromosomes, existing as 23 pairs, while germ cells have only one set. Alleles of genes in the same linkage group, i.e. linked together on the same chromosome, tend to stay together more often than alleles of genes in different chromosomes. But even linked genes tend to be shuffled or *recombined*, as the chromosomes pair up and exchange parts in the formation of germ cells. So, if one of the chromosomes in the pair contains the sequence of genes, *ABCDEFG* and its partner, (the homologue), *abcdefg*, they may exchange parts to give *ABcdefg* and *abCDEFG* or other possibilities, which usually, though not always, preserve the order in which the genes occur on the chromosome. The more closely-linked two genes are, the less likely it is that the linkage will be disrupted. For example, *A* and *B* are linked right next to each other, so it is less likely for them to get unlinked, than *A* and *C*, or *A* and *D*, and so on. Similarly, the linkage between *a* and *b* will be less likely to be disrupted than between *a* and *c*, or between *a* and *d* in its partner. In fact, by looking at the frequency with which linked alleles become unlinked or recombined, it is possible to estimate how close they are located to each other on the chromosome. This is the basis of much of the gene hunting that currently goes on, as you will see in Chapter 12.

Mendel's laws are typical of western scientific theories, of which the archetype is Newtonian mechanics. One fixes on the essential variables - the ratios of phenotype classes in the case of genetics and different kinds of motion in the case of mechanics - and proposes to explain how they come about in terms of the behaviour of (invisible) controlling entities. In the case of Newton's laws, it is force, or gravity; in the case of Mendel's laws, it is the genes. In so doing, Newton has banished the vibrant world of colour and form, of light and music, replacing it with a silent universe of lifeless, immobile objects subject to the push and pull of external forces. In the same way, Mendel has abstracted away the untidy complexities of living, experiencing organisms for ideal, eternal entities that behave in a simple, logical way.

There is nothing wrong in abstraction and idealisation, provided one does not thereby lose sight of the reality. However, the unfortunate tendency is to mistake the abstraction for reality - a case of the fallacy of misplaced concreteness - which is where *reductionism* begins. People start to forget the organism altogether by thinking of it as a mere collection of genes. Thus, they discount not only how genes can give rise to organisms (if indeed they can[2]), but also how they interact with one another and are absolutely dependent on the cellular and ecological contexts. The focus on genes has eclipsed the physiological as well the physicochemical and mathematical contributions to our understanding of living processes and organisation.[3]

Where did Mendel's Theory come from?

How did Gregor Mendel come by his theory? Did he dream it up out of nothing? Was it simply the result of years of painstaking experimentation, of growing thousands of pea-plants and scoring characters in tens of thousands of peas? It will be of interest to examine how Mendel arrived at his theory which, together with Darwin's theory of evolution, became the twin pillars of the genetic determinist paradigm.

People of all ages the world over have thought about heredity in connection with their own reproduction and kinship. One of the earliest written records in the west comes from Lucretius, who describes particulate inheritance as an explanation of why characters can skip generations.[4] Similarly, selective breeding of domesticated plants and animals has been carried out by farmers everywhere for at least tens of thousands of years. So, what gave Mendel the edge, and was Mendel's theory so special after all?

Mendel came from a peasant farming family in a small village in Moravia, where horse-breeding was a major preoccupation. Mendel excelled in school work and became an Augustinian monk in Brno. While there, he worked as a supply teacher in the local high school, and enrolled as a student in the University of Vienna. He acquired a good grounding in physics and chemistry, as well as biology. Mendel was thus predisposed to apply the concepts and quantitative methods of physics to biology: mechanics, atomic theory and, especially, statistical mechanics - the mechanical behaviour of matter in terms of large numbers of atoms or molecules.[5]

Another important influence was the idea of evolution, which had been increasingly debated in learned circles in Europe. Mendel's biology teacher at Vienna University, Franz Ungar, was a supporter of evolution. He investigated the process of fertilisation in flowering plants, and knew a great deal about earlier hybridisation experiments which had been carried out since the latter half of the eighteenth century. Ungar emphatically rejected the belief taught by the Christian Church that species were created and fixed once and for all. Instead, he held that variants arose naturally in the wild, some of them sufficiently different to become new species.

The question of whether or not animal and plant species had been fixed and immutable since the beginning of creation was indeed one which captured the imagination and passion of the rising middle-classes in Europe, who were becoming more and more independent of the Christian Church on the one hand, and the aristocracy on the other. This period also saw major advances in mechanical physics which were to serve as a model for scientific theory in all other disciplines.

The early hybridisation experiments, described by Mendel's teacher, Ungar, were pioneered by Joseph Koelreuter, Professor of Natural History in Karlsruhe in Germany in 1760, who carried out 500 different hybridisations involving 138 species, with parallel investigations on the mechanisms of pollination and fertilisation. The work lay neglected until repeated and confirmed by Carl von Gaertner. Both Koelreuter and Gaertner concluded, on the basis of their experiments, that artificial hybrids, even when they succeeded in setting seed and developing into plants, did not breed true, and hence, by implication, species were the only natural entities which were fixed from the beginning. Furthermore, they concluded that hybridisation was of no importance at all in giving rise to new species. Mendel's teacher Ungar, however, came to just the opposite conclusion on the basis of the same experimental results. There is no necessary connection, in scientists' minds, between observations and the theory itself (as has been made so abundantly clear in the current genetic engineering mindset that totally ignores or misreads observations). Scientists, like all human beings, are strongly influenced by their preconceived notions of what seems right. Thus, Koelreuter and Gaertner were still strongly influenced by the idea that the world was created by God once and for all eternity. Ungar, instead, had already rejected this static world view for an evolutionary perspective in which change and transformation took centre stage.

Mendel decided to settle the issue of whether species were fixed or whether hybridisation could give rise to new species by carrying out more experiments. We can see that ideas about evolution and heredity were intimately associated right from the start.

The Questionable Mendelian Ratios

Both Koelreuter and Gaertner were prodigious and diligent experimenters and had noticed that when two varieties were hybridised, the offspring, or the first filial generation, F1, were uniformly alike, and intermediate between the parental varieties in many characters. When the F1 hybrids were crossed among themselves or self-pollinated, however, the resulting F2 generation threw up a great many variations and only some characters gave ratios approximating the 3:1 or 1:2:1 that Mendel later observed. (Based on what you have learned earlier in this chapter, you can verify that those characters were homozygous in the parental varieties, and heterozygous in the F1.) Neither, however, succeeded in formalising the results as Mendel did. One important reason was that *not all characters behaved in that way*. This was also the case with Mendel's own experimental findings. We now know that *Mendelian heredity applies to only a restricted range of morphological characters, observed over a limited number of generations under more or less unchanging environmental conditions, and only when the organisms within the same species, with the same number of chromosomes, are crossed.* It was found, not so long ago, that one of the characters Mendel studied, wrinkled peas, were actually brought about by a mobile genetic element, or a jumping gene, that had inserted itself into the normal wild-type allele for round peas - a definitely non-Mendelian process.[6]

Mendel's laws are a great simplification and idealisation. Furthermore, they did not come from his observations, nor from the observation of his predecessors. Where did they come from?

The clue is to be found in one of Mendel's letters to the botanist Carl Naegeli, in which he describes his own task to ascertain the 'statistical relations' of the different forms among the offspring of the hybrids. Mendel had obtained a good knowledge of mathematics from his studies in physics. Christian Doppler and Andreas von Ettinghausen were his physics teachers, both of whom emphasised the mathematical approach. There were also correspondences between the Abbot of the monastery in Brno and Carl Friedrich Gauss, a founding father of probability theory and much else besides, who was concerned with the errors of measurements. The Abbot himself took a great interest in Mendel's work, and in all probability Mendel knew what to look for in his experiments. In other words, he had proceeded deductively from general principles to specifics, with a mathematical theory already in mind which he sought to confirm by experiments.

The statistician, Ronald Fisher, a prominent figure in the development of population genetics and neo-Darwinian theory of evolution, analysed Mendel's results statistically, and concluded that the agreement with the 3:1 ratio was so perfect that 95 times out of a hundred the result could not have been obtained. In other words, the agreement with expectation based on the gene theory was too good to be true. Did Mendel fabricate his results? Not at all. As historian of science, Robert Olby, points out,[7] Mendel merely stopped scoring when he got the right numbers, because he knew what ratio he was looking for. Who can blame him? Scoring peas, as such, is a tedious task, particularly if you know the answer already.

As mentioned earlier, there were many characters that did not fit the neat pattern of Mendelian heredity. But Mendel only concentrated on those that did. The resulting theory therefore can give, at best, a very partial and idealised account of heredity. The parallel between Mendel's laws of heredity and Newton's laws of mechanics and statistical mechanics is more than coincidental: the one is modelled on the other. Newtonian mechanics was supplanted at the beginning of the present century by Einstein's theory of relativity on the large scale, and by quantum mechanics in the sub-microscopic domain. Mendelian genetics has suffered a similar fate since the 1970s, but not before it had inspired a great number of discoveries which culminated in the DNA double helix and the cracking of the genetic code.

One lesson that can be learned from this episode is the provisional nature of all scientific theories. What is more important, I believe, is that people should not take scientific theories too seriously by treating them as 'laws' of nature, as though they had been laid down by God. Scientific theories, and mathematics, for that matter, are tools for helping us think more clearly in our effort to understand nature, and in our quest for the poetry in nature that is ever beyond what words or theories can say.[8]

Notes

1. A man actually gets more genes from his mother. Genes occur on chromosomes, which, like the genes they carry, occur in pairs. Humans have 23 pairs of chromosomes, the members of each pair are alike, of the same size (and differ from those of other pairs), except for the sex-determining chromosomes. Females have an equal pair, XX which are both large; males, on the other hand, have an unequal pair, XY, the Y chromosome being a lot smaller. The father would have inherited his Y chromosome from his father and the X chromosome from his mother.

2. The role of genes in development is predominantly to provide the enzymes for metabolism and the material for making the organism. Another role might be to stabilise patterns generated by dynamic physicochemical processes. This view has been promoted by a number of authors: see Ho and Saunders, 1979; Webster and Goodwin, 1982, 1996; Ho, 1984b; Saunders, 1984; Goodwin, 1984.

3. A number of recent volumes are attempting to redress this balance. See Ho, 1993. Goodwin, 1994; Laszlo, 1994; 1996; Saunders, 1997.

4. Saunders, 1998.

5. Olby, 1966.

6. Battacharyya *et al*, 1990.

7. See Olby, 1966.

8. See Ho, 1993

Chapter 6

Neo-Darwinism - Triumph or Travesty?

Neo-Darwinism is the marriage of Mendelian genetics and Darwinism. The theory of the gene was based on evidence that was directly in conflict with it, only the climate of biological opinion was favourable. Neo-Darwinism became established as a theory that purports to explain everything, and is thereby in danger of explaining nothing. A theory that lacks content is easily pressed into the service of pernicious ideologies.

The Theory that Explains Everything

The marriage of Mendelian genetics with Darwinism was not an immediately happy affair. In this chapter I shall briefly review the history of how neo-Darwinism became established as a dominant theory that purports to explain every aspect of the organism, yet ends up eclipsing the organism totally from view. It is important to examine in some detail the anatomy of a neo-Darwinian explanation to realise why the theory is so easily pressed into the service of pernicious ideologies.

Gradual Evolution versus Evolution in Jumps

As soon as Darwin's theory of natural selection was proposed, argument began as to whether small or large variations were important for evolution. Those who believed in gradual evolution concentrated on characters that vary continuously in the population, such as height, body weight, milk yield, and so on. Others who opted for large changes, however, focused on characters that vary discontinuously, i.e. those that fall into a few discrete or non-overlapping classes, such as the colour or shape of flowers or the presence or absence of a structure.

The question of whether continuous variations or discontinuous variations are the stuff of evolution began with Darwin's insistence that "*natura non facit saltum*" (nature does not make jumps); natural selection acts on the "insensibly fine gradations" of continuously varying characters. Darwin remained firmly wedded to gradualism even though his staunchest supporters, Thomas Huxley and Francis Galton, both pointed out that large, discontinuous variations could also be subject to natural selection and, if anything, made it easier to produce evolutionary change.

Consistent with his concept of gradual evolution, Darwin believed in blending inheritance. His theory of *pangenesis* proposed that during the lifetime of individual organisms, their organs and tissues throw off infinitesimal 'gemmules' or 'pangenes' into the bloodstream and that these are transported to the germ-cells, thus determining the characteristics of the next generation. Those organs that are more heavily used will contribute proportionately more influence; conversely, those that are disused will be underrepresented. The theory predicts that the influences of successive generations are continually being mixed and blended and cannot account for discontinuous changes, or for characters that miss generations, both of which require a particulate theory, as we saw in chapter 4.

Francis Galton argued that Darwin's theory of pangenesis would be ineffective in changing the species under natural selection, as the offspring generation would merely regress to the mean of the parental generations, and the influence would, in any case, be increasingly diluted out in successive generations. He also showed that when rabbits were transfused with blood from different varieties, they did not produce offspring with characters intermediate between those of the blood donor and the recipient. He concluded, therefore, that continuous variations could not be the basis of evolutionary change. On the contrary, evolution must occur in jumps, through the appearance of discontinuous variations.

Discontinuity versus Continuity[1]

William Bateson had been studying natural as well as artificially induced variations in plants and animals which often fell into discrete, non-overlapping classes. Furthermore, many species showed parallel variations. Bateson was trying to work out how those characters were inherited when he discovered Mendel's paper. By that time, he had already been in dispute with the biometricians, Karl Pearson and W.F.R.Weldon, who, working on Darwin's hypothesis of blending inheritance, also believed that continuous variations were the raw material for natural selection. (Biometricians use mathematical techniques to study continuous variations in populations.) Their disagreement soon escalated into a full debate in which others joined in, with the Mendelians on one side and biometricians on the other. This debate was eventually resolved by the birth of population genetics. At least that is how the story is told.

Pearson and Weldon, like Darwin's cousin Francis Galton before them, used statistical techniques to study continuous variations. They refined and corrected Galton's mathematical derivations, showing how, contrary to Galton's claim, natural selection could act on continuous variation if the extremes were selected for breeding, in which case, the offspring would not regress to the mean of the parental generation, as Galton had claimed.

In opposition to Pearson and Weldon, William Bateson did indeed insist that discontinuous variations are the stuff of evolution and, on discovering Mendel's work, believed that the latter had provided a theory of the inheritance of discontinuous variation. But he was by no means satisfied by the explanation of evolution in terms of the natural selection of Mendelian genes. It was his belief that no explanation of evolution is complete without a theory of how Mendelian genes could produce the variation of biological form. It was clear to him that organisms do not vary at random, but rather, in accordance with certain 'laws' or regularities governing growth and variation. This, incidentally, forms much of the substance of the present-day alternative approaches to neo-Darwinism.[2]

Bateson's debate with Pearson and Weldon was as much a conflict of personalities as a power struggle over the control of the 'Evolution Committee' of the Royal Society, the trend-setting Society of mainstream scientists established by King Charles II in the 17th century that has dominated British science up to the present day.

In my opinion, the important disagreement between Bateson and the biometricians was not over whether continuous or discontinuous variations are the stuff of evolution. It was over how evolution should be explained. Pearson and Weldon were content to explain evolution in terms of hereditary influences that pass from one generation to the next and that account for correlations between generations. The reason they concentrated on continuous variations was because they were mathematically tractable using the linear, additive models that allowed equations to be solved. They were not interested in the real biology of organisms. In contrast, Bateson was dissatisfied with explanations based on abstract hereditary influences or, for that matter, on Mendelian genes. In that respect, he was *not* a Mendelian, for he was after a more fundamental explanation of biological forms and variations; of how they are *generated* during development. For example, he suggested that discontinuous variations correspond to points of stable equilibrium in a continuum of possible forms, and that repeated body parts or segments - common to many groups of organisms - are generated by physical mechanisms similar to standing waves or vibrations. These ideas are borne out by a large body of contemporary work that attempts to understand morphogenesis in terms of non-linear physicochemical, mathematical models.[3]

Eventually, Mendelian genetics became united with biometrics to give biometrical and population genetics, in which continuous variations are interpreted in terms of the segregation of many genes (i.e., polygenes or multigenes), each giving a very small effect that are all added together. This is usually said to be the resolution of the debate, but is really a major digression. The question on the real causes of morphogenesis and evolution has never been settled or resolved in the exclusive emphasis on genes in mainstream evolutionary theory ever since.

The Myth of 'Pure Lines' and the Gene Theory

A crucial step in the development of the gene theory was taken by Swedish botanist, Johannsen, who coined the terms 'gene', 'genotype' and 'phenotype'. He carried out 'pure line' experiments, which purported to show that a genetically uniform line bred true to the genes it carried (its genotype), and did not respond to selection of visible characters (its phenotype). Thus, in a genetically uniform line of self-fertilising French beans, he bred from the smallest and the largest seeds, and found that the offspring gave the same range of variation in seed size. In other words, there was no correlation or resemblance between the size of the parental seed and the size of the seeds in the offspring. Actually, Johannsen found statistically *significant* correlation between parent and offspring[4] which he chose to ignore, and instead, *interpreted* his results in accordance with Weismann's theory of the germplasm. Several other scientists soon followed suit in demonstrating how selection was ineffective in 'pure lines', all based on the *a priori* acceptance of Weismann's theory and not on actual experimental observations. This was eventually soundly criticised by the American biometrician, J. Arthur Harris, who, on reviewing the entire body of work, pinpointed the real reason why the 'pure line' theory was so readily accepted, based on evidence that was directly in conflict with it: the climate of biological opinion was favourable to pure line theory.

Thus, the experimenters' reasoning was circular. The continuity of the germplasm was assumed, they then observed that selection could not change the pure line. If it did, the line was not pure. And, on that basis, it was concluded that selection had no effect on pure lines. As the genotype was, then, inaccessible to direct observation, there was no independent measure of genetic uniformity or heterogeneity. Similar circular reasoning soon became the established practice of the genetic determinist paradigm, especially as the subsequent mathematicisation of the gene theory in population and biometrical genetics accorded it the status of a rigorous science.

The Mathematicisation of the Gene Theory

Mendel's theory already had a mathematical structure that lent itself to generalisation. Thus, it was only a matter of time before it was extended to describe how genes behave in populations. This step was first taken in 1902 by Yule, who showed that the Mendelian ratios 1AA: 2Aa :1aa in the second generation after hybridisation remains unchanged in subsequent generations, *so long as individuals in the population mate at random.* This means purely according to chance, so that matings between the same or different genotypes are not more or less frequent than their frequencies in the population would predict. Moreover, the frequency of both alleles also remains constant in all subsequent generations. This was further generalised later, independently, by British mathematician Hardy, and German physician Weinberg, who showed that in large, randomly mating populations, any frequencies of two alternative alleles of a gene, say, *A* and *a*, will remain constant in the absence of selection. This is the Hardy-Weinberg 'law', which states that if the frequency of the *A* allele is *p* and that of the *a* allele is *q*, then the proportion of genotypes, *AA*, *Aa* and *aa* are given by the binomial expansion, $(p+q)^2 = p^2 + 2pq + q^2$, and will so remain for all subsequent generations. In other words, it takes only one generation to reach an equilibrium - the Hardy-Weinberg equilibrium. (However, when there are more than two alleles segregating, it takes more than one generation to reach equilibrium, as Weinberg showed later on.) The Hardy-Weinberg law forms the basis of population genetics later developed by Fisher, Haldane and Sewell Wright, though each of them emphasised different aspects.

The debate between the Mendelians and biometricians was, as I said, a distraction. As soon as particulate inheritance was accepted, it was possible to interpret continuous variations as the result of many genes segregating, each with a small additive effect. This was first suggested by Yule in 1906, in the midst of the continuing debate between Bateson and Pearson.

By 1918, mathematical treatments of the effects of selection in Mendelian populations had already been done. The ground was prepared for Darwinism to be reinterpreted according to the gene theory in the 'neo-Darwinian synthesis' from around the 1930s up to the 1950s and 60s.

At the base of the neo-Darwinian synthesis is the mathematical representation of genes in population by Ronald Fisher, J.B.S. Haldane, and Sewell Wright, all of whom were concerned to show that Mendelian genetics and Darwin's theory of natural selection were not in contradiction, but could be unified into a single, rigorous mathematical theory. Fisher's approach was to concentrate on selection of *polygenes* - many genes each with a small effect - in large populations over long periods of time. This was closest to Darwin's original conception of gradual evolution. Fisher formalised this scheme to 'the fundamental theorem of natural selection' which states that, "The rate of increase in fitness of any population is equal to the genetic variance (variation) in fitness". This theorem assumes that genes have small, *additive* (non-interacting) effects on the character under selection. In other words, it assumes that the genes act independently of one another. Based on this model, Fisher worked out estimates concerning the *heritability* of traits, which refers to the proportion of the observed variability in a population that can be attributed to or associated with the underlying genetic variation. This is the estimate that is most frequently used today by biometrical geneticists in controversial studies on such measures as IQ scores.

Sewell Wright, by contrast, recognised the importance of *epistasis* or interactions between genes at the very beginning of his career through selection experiments on guinea-pigs in the laboratory of the biologist, William Castle. He found that the inheritance of coat colour was controlled by a system of genes that interacted non-linearly with one another, rather than acting additively and independently. This convinced him that selection acted on whole *combinations* of genes and not on single genes. His picture of evolution involved selection acting on chance combinations of genes giving large effects, which are fixed in small populations, rather than on individual genes with small effects in large populations. He recognised that the same genes will have different effects in different genetic backgrounds, and developed the most rigorous mathematical theory of path coefficients that enabled experimenters to make inferences concerning environmental *versus* genetic effects, as well as additive and non-additive, interactive effects. However, in order to perform analyses by path coefficients, one has to have a huge amount of data as well as knowledge concerning the breeding habits of the population, i.e. whether mating occurs at random or whether inbreeding and other non-random matings are practised. These requirements are almost never satisfied except in controlled plant and animal breeding experiments.

J.B.S. Haldane's theory is intermediate between that of Sewell Wright and Fisher. Although he concentrated on the selection of single genes with large effects, he did occasionally recognise the importance of dominance and other interactions between genes. Overall, his theory, like that of Fisher, fits the description of 'beanbag genetics', a term of derision used by the biologist Ernst Mayr, who was extremely critical of mathematical approaches to evolution in general which did not take into account the non-linear interactions between combinations of genes. However, Mayr's own insistence that genes acted together as 'co-adapted gene complexes' would seem to place him fairly close to Sewell Wright's position. The main difference seems to be that Mayr's co-adapted gene complexes are specifically selected for, so that they remain associated together more often than can be accounted for by chance. However, he never proposed any mechanism as to how that could be achieved, given that linked genes are also recombined at random during reproduction, according to the theory.

The mathematicisation of Darwin's theory in terms of the natural selection of genes (determining characters) was the important turning point for the general acceptance of the theory, for it accorded it a status equal to Newtonian mechanics, statistical mechanics, or the second law of thermodynamics. The mathematics gloss over the fact that, while the theory says a great deal about how genes behave in populations, it offers *no* real explanation of how organisms can develop or evolve.

A neo-Darwinian explanation typically starts by identifying a character which is assumed to be controlled by a gene that confers a selective advantage or disadvantage, so that it is selected for, or against; and that explains why the organism does or does not possess the said character. In this way, it is possible to 'explain' any and every character that the organism is said to possess.

However, the link between genes and 'characters' is, in most cases, not at all straightforward, and the problem has not been solved by identifying genes that affect the characters. The most concrete thing that is known about genes, as you will see in Chapter 7, is that they code for proteins, or for signals that regulate the synthesis of different proteins. It is a big conceptual jump from that to the characters of organisms.

An immediate difficulty arises in the identification of the character itself. It is one thing to name a character such as hair colour, or colour of eyes. It is quite another to say that there is a character called 'aggression', for example. Animals may engage in aggressive acts, but that does not mean there is a character called aggression. Similarly, some humans may show preference for the same sex, that does not mean there is a character called homosexuality. Both are social acts carried out in certain contexts. To invent a corresponding character trait and, on top of that, a gene determining it, is to commit the fallacy of reification - mistakenly creating *things* for processes. There may be many mutations in many genes which affect a person's ability to read or speak or remember things, but that does not mean those are genes *for* reading, speech or memory. Even in the case of morphology, there is no theoretical or conceptual basis to justify the separation of a 'character' from the interconnected whole that is the organism. But that is effectively what is being done by geneticists who study development. They discover mutations in some gene that affect some aspect of development - say, the segmentation pattern of the body - and that gene becomes a 'gene for segmentation', or 'segmentation gene'.

The second difficulty, which follows from the first, is the equally unrealistic assumption that there are polygenes determining or influencing the supposed character, and that they act in an additive way, i.e. each independently of all the rest. For this goes against everything that is known about gene action and metabolism (see Chapter 3). Genes are involved in the development of all aspects of the organism. But the individual genes cannot be extricated from the whole context of the organism in its eco-social environment.

Neo-Darwinian explanations, in purporting to explain everything, ultimately explain nothing because there is no independent verification of the 'adaptive story' which must be invented to 'explain' how the character is selected for, or against. Nevertheless, this completes the circle which validates and legitimises the hunt for the genes or genetic basis of the characters in question. The inherent danger in this kind of reasoning is that it is all too easy to reinforce the prejudice with which one begins, giving rein to the worst excesses in eugenic, racist ideologies in the present century. Let us look at a key concept in biometrical genetics, *heritability*. This continues to be misused by the genetic determinists of our day, in their attempt to lend credence to the existence of dubious characters such as intelligence or criminality and then to use the supposed measure of such characters to stigmatise individuals, races or social classes.

Heritability of IQ, and other polygenic Traits

The public are often told that there is a large genetic component to intelligence as measured by IQ or to criminality, as measured by some psychometric tests. It is often claimed that such character traits are 50% due to the genes. These statements are based on an erroneous interpretation of estimates on heritability from biometrical genetics.

The first thing one has to realise about heritability is that, contrary to what is claimed, it does not tell us anything about the degree to which a character is determined by the genes. Technically, it is measured as the proportion of the total (phenotypic) variation in a *population* which is due to genetic variation:

$$\text{heritability} = (\text{genetic variation})/(\text{total variation})$$

The total variation is assumed to be made up of those due to the genes plus those due to the environment, so:

$$\text{heritability} = \frac{\text{genetic variation}}{\text{genetic variation + environmental variation}}$$

Heritability is a population measure and says nothing about the individual.[5] Furthermore, it is specific for the population only of that particular *generation*, because the environmental conditions may differ for different generations. Estimates of heritability carried out on the same plant varieties have been found to vary widely in successive years. That alone should convince us that heritability is not a constant property of any trait, whatever. It is therefore pointless to measure heritability unless the population is being bred continuously in an absolutely constant, controlled set of environmental conditions.

In order to measure heritabilities rigorously, a very laborious breeding programme has to be set up. It goes like this. One has to extract a dozen or more 'pure lines' (genetically uniform lines) from the population by many successive generations of inbreeding until, theoretically, no more genetic variation remains. If the lines survive the inbreeding, and many of them may not due to inbreeding depression, then they will be homozygous in most if not all of their genes. Next, the lines have to be crossed in pairs to produce the F1 generations, which will all still be genetically uniform although all individuals will be heterozygous for all their genes instead of being homozygous as in the 'pure lines'. The variation of the character in question can then be compared for the pure lines, the F1 hybrids and the original 'random mating' line in a suitably randomised experimental design, so that the range of environments experienced by each line will be the same. Now, according to the theory of biometrical genetics, the variance (a mathematical measure of variation) in each of the pure lines as well as the F1 hybrids are due to the environment alone, as there is no genetic variation between individuals. So the average of their variances gives an estimate of the environmental component of the variation. The variation in the random mating line, on the other hand, will consist of both the genetic and environmental variation. The important point about this estimate is that the genes in question have to be randomised over all environments and over all genetic backgrounds. This means that the population needs to be very large and mating has to take place strictly according to chance (i.e. *randomly*) and not more or less frequently than that predicted from the frequencies of the genotypes.

Now, if the population is, indeed, randomly mating and very large, then some shortcuts are available. One can estimate heritabilities by how much offspring resemble their parents, or siblings (brothers and sisters) resemble each other, or ultimately, how much monzygotic twins (i.e., twins derived from a single egg, and hence completely identical in their genetic makeup) are alike. The difficulty is that human populations are far from random-mating, and the genes involved are, therefore, not at all randomised with respect to the genetic background. The stratification of the environment by class and other social factors also means that the genes are far from randomised over all environments. This makes heritability estimates of human polygenic traits very unreliable, even in cases where one assumes it is meaningful to treat as 'traits' such measures as IQ scores. Much is made of the strong resemblance that is commonly found between identical twins who have been separated from birth. This is used to support the idea that certain traits have a very high 'genetic component'. These twin studies actually suffer most from the limitations imposed on them by small samples, with highly non-random association of environments and of genetic backgrounds. We shall look at the whole issue of IQ measures in more detail in Chapter 12.

The most reasonable statement one could make with regard to all characteristics of the organism is that genetic and environmental factors are *inextricably intertwined*. This is borne out by everything we know about the molecular basis of gene function (see Chapter 8). Should anyone think of attempting these classical breeding experiments to measure heritability, don't. They are pointless, for it is actually impossible to derive really pure lines. 'Pure lines' are among the purest genetic fictions ever invented. Anyone who has experience of keeping genetic stocks in the laboratory knows that they have to be periodically 'purified', otherwise they become very variable. We now know that, on account of the fluidity of the genome, new variations will keep arising in the different lines, even as they are being bred to be homozygous and uniform. Moreover, it is well-known from classical genetic analyses that populations which are more genetically uniform tend to be phenotypically *more* variable. This is attributed to the failure of 'developmental homeostasis' - the totality of regulatory mechanisms of the system that buffer it against perturbations, keeping it constant and stable.

Notes

1. See Ho, 1997b for a recent review on evolution.

2. An important element in recent alternative approaches to evolution is the emphasis on the dynamics of developmental processes which generate non-random forms. So much so that a *rational taxonomy* of the forms could be derived; see Ho and Saunders, 1979; 1984; 1993; 1994; Goodwin, 1984; 1994; Webster and Goodwin, 1982; 1996; Saunders, 1984; 1997; Saunders and Ho, 1995; Ho, 1984b, 1988b; 1990; 1992.

3. See for example, Turing, 1952; also Saunders, 1984; 1998; Webster and Goodwin, 1997, and references therein.

4. The history of population genetics is told by Provine, 1971.

5. For an authoritative, in-depth critique of heritability estimates, see Lewontin, 1982. The treatment here differs in detail, though not in substance. Richard Lewontin has published extensively on the IQ debate and related issues.

Chapter 7

The Central Dogma of Genetic Determinism

In the years following the discovery of the double helix, the organism has become totally eclipsed. It is seen as a collection of genes which control development according to a 'genetic programme'. Evolution is said to occur by the natural selection of random mutations, ensuring that the fittest mutants survive to reproduce. Thus, the organism is projected as a passive object, buffeted by selective forces over which it has no control. Its life experience is negated, as only the genes it carries are of any evolutionary consequence. It is no accident that a culture bent on promoting capitalism and free enterprise should be obsessed with things rather than processes. The notions of 'gene banks' and 'genetic resources' make it plain that life, the process of being alive, as well as real organisms and diverse ecological communities, are all negated in favour of genes that can be grasped hold of, possessed, preserved and exploited as commodities.

From Theory to Molecules

The neo-Darwinian synthesis was predominantly a theoretical exercise involving mathematicians dealing with abstract hereditary factors and experimenters who were primarily interested not in what genes were, or what they did, but in how breeding experiments could be conceptualised in terms of hereditary influences or factors. By contrast, molecular biology developed out of a deliberate search for the material basis of heredity which was initiated by the German biologist, August Weismann. Weismann was responsible for proposing the theory of the continuity of the germ plasm, which, as you have seen, exerted a great influence over the neo-Darwinian synthesis.

The Immortal Germplasm and the Soul

The theory of the germplasm depended on two key observations.

First, in insect embryos, Weismann found that the future germ cells became separated from the rest of the body very early in development. From this he deduced that germ cells are protected from environmental influences, so that the *germplasm* they carry retains its constancy and immortality. Second, he discovered, in the fertilised egg of the round worm, *Ascaris*, a pair of large chromosomes, whereas the unfertilised egg had only one of them. From this, he deduced that the germplasm resides in chromosomes, whose number is halved in the germ cells, so that when male and female germ cells unite on fertilisation, the original number is restored. Weismann's theory of the continuity and immortality of the germplasm was idealism through and through. As many later commentators pointed out, the early separation of germ cells is peculiar to insects; it does not occur generally in animals, and not at all in plants. Furthermore, *there is no evidence that the genetic material in germ cells is immune from environmental influence* (see Chapter 8). As mentioned in Chapter 4, the germplasm took on the role of the immortal soul when science replaced the Christian religion. Nevertheless, Weismann had correctly identified the chromosomes as the bearer of genetic material, which immediately narrowed the search for the material basis of entities that behaved like Mendelian genes. The occurrence of chromosomes in pairs and their segregation (separation) in the germ cells exactly paralleled the segregation of Mendelian genes; moreover, chromosomes, just like genes, were also subject to independent assortment in the formation of germ cells, so that each germ cell ended up with a different combination of maternally-derived or paternally-derived chromosomes (see Chapter 4).

The Chromosome Theory of Inheritance

Further advances were made from about 1910, when Thomas Hunt Morgan set up a laboratory devoted to studying the chromosomal genetics of the fruit fly, *Drosophila melanogaster*. He and his colleagues eventually established that genes were linked together in linear arrays, each array corresponding to a chromosome. Each chromosome contained thousands of genes. To prove that genes are linked together in linear arrays on chromosomes, large numbers of mutations were produced with X-rays. Lines homozygous for different alleles in different genes were established and crossed with one another so that the progeny could be analysed for the genes deviating from independent assortment. It was found that genes located on different chromosomes would assort independently, whereas those on the same chromosome would not. And the more close together two genes were on the same chromosome, the less likely it was that the linkage would be disrupted. In this way, genes were assigned to chromosomes and detailed genetic maps of chromosomes produced. As microscopic techniques improved, impressive changes in chromosomes corresponding to various changes in the linkage maps of genes could also be identified. The resulting 'chromosomal theory of inheritance' may be summarised as follows:

The chromosomal theory of inheritance

1. *Genes occur in linear arrays on chromosomes.*
2. *Different chromosomes contain different combinations of alleles of genes.*
3. *Chromosomes occur in (homologous) pairs in sexually reproducing organisms (such as human beings).*
4. *Only one of each pair of chromosomes is present in germ cells.*
5. *In the formation of germ cells, homologous chromosomes exchange parts so that the linked genes are recombined: The more close together the genes occur on the chromosome, the less likely they are to be recombined.*

Morgan started out as a developmental biologist extremely sceptical of the chromosomal theory, for he rightly regarded the process of heredity as inseparable from that of development. The subsequent success of the chromosomal theory, together with Weismann's theory of the germplasm, however, was precisely to separate the two. Thus, the process of development was no longer considered important for the study of evolution, which consisted exclusively in the natural selection of random mutations. As Morgan stated in 1916,

"If, through a mutation, a character appears that is neither advantageous nor disadvantageous, but indifferent, the chance that it may become established in the race...is extremely small... If, through a mutation, a character appears that has an injurious effect,... it has practically no chance of becoming established.

"If, through a mutation, a character appears that has a beneficial influence on the individual, the chance that the individual will survive is increased, not only for itself, but for all of its descendants... It is this increase...that might have an influence on the course of evolution."[1]

This is essentially the 'beanbag' theory of evolution that most population geneticists came to embrace.

The DNA Double Helix and the 'Central Dogma' [2]

The chromosomal theory established the material basis of genes. But what did the genes actually do? As long ago as 1902, the physician Archibald Garrod recognised that 'inborn errors of metabolism' in humans were metabolic blocks produced by defects in single genes. One of the main functions of genes therefore, was to control specific chemical reactions in metabolism. The detailed analysis of that was carried out by George Beadle and Edward Tatum in the bread mould *Neurospora crassa*. They showed that each gene is responsible for the synthesis of one enzyme; and a mutation in each enzyme leads to a specific metabolic block. Enzymes are proteins, which are long polymers made up of hundreds of simpler units called amino acids. There are twenty different amino acids, all of which have the general chemical formula, NH_2-CH(R)-COOH, and differing from one another in the side-chain, R. (see Fig. 7.1). There is a great diversity of proteins each of which has a specific sequence of amino acids. For example, for a protein with 100 amino acids, the number of possible sequences is 20^{100}.

For a long time, there was a debate as to whether the genes on chromosomes were actually the enzyme proteins themselves or the nucleic acid DNA, which was also known to be a chemical constituent of the chromosomes. This was resolved by Avery, MacLeod and McCarty, who were studying the bacterium, *Streptococcus pneumoniae*, frequently isolated from patients with pneumonia. The bacterium exists in two forms: a smooth form which causes disease, and a less virulent rough form. The experimenters showed that, when DNA was extracted from the smooth bacteria and added to cells of the rough type, a high proportion of the latter were *transformed* to the smooth type. Transformation by direct uptake of DNA is one of the regular routes whereby different bacteria can exchange genes. We shall look more closely at gene transfers between micro-organisms in Chapter 10.

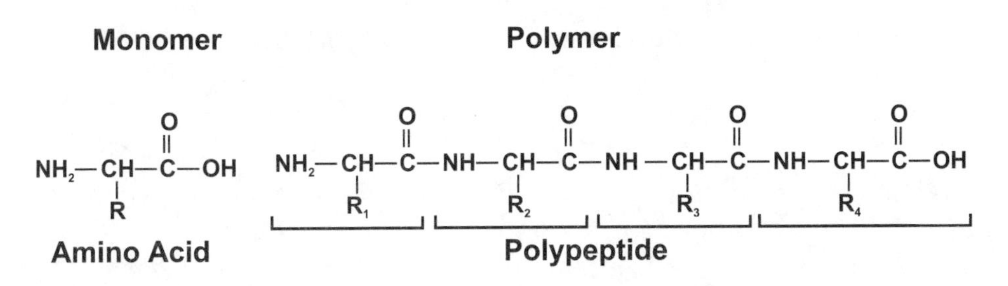

Figure 7.1 Chemical structure of proteins.

Bacterial cells, like all other cells, are susceptible to infection by viruses. Almost all infectious viruses contain DNA or a related nucleic acid, RNA, which are also found in all cells. Once in the cell, the DNA or RNA can take over the cellular machinery to make more viruses. Viruses have protein coats around a core of nucleic acids. When different strains of viruses infect the same cell, they can exchange genes to make new viruses. In 1952, eight years after the experiment of Avery, MacLeod and McCarty, Alfred Hershey and Martha Chase showed that it was sufficient to have only the DNA of the virus, without its protein coat, to cause the infection of the bacterial cell.

DNA, or deoxyribonucleic acid, was also known to made up of very long polymers consisting of simpler units called nucleotides (Fig. 7.2). Each nucleotide is made of a sugar, deoxyribose, to which is attached an inorganic phosphate group and an organic base. The nucleotides differ in the organic base they contain, of which there are only four, adenine, guanine, cytosine and thymine, (A, G, C, T, for short). Another kind of nucleic acid which is abundant in the cell is RNA, or ribonucleic acid. This is very similar to DNA except that the sugar is ribose instead of deoxyribose, and the base uracil U replaces thymine T. Each nucleic acid polymer is made of many nucleotides joined together by phosphodiester bonds between the sugar of one nucleotide and the phosphate of the next, so that a backbone of alternating sugar and phosphate groups is formed, with the bases sticking out on the side. Apart from that, little was known about the 3-dimensional structure of DNA, which is important as that holds the key to how it functions as genetic material.

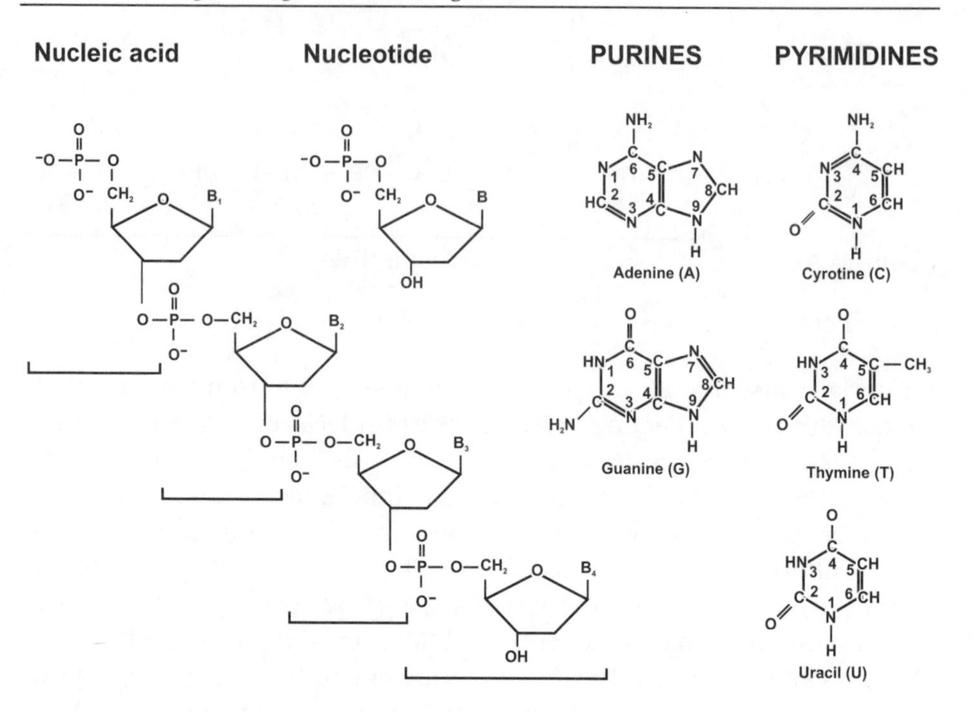

Figure 7.2 Chemical structure of DNA and RNA

At the time, Linus Pauling had worked out the 3-dimensional structure of proteins by X-ray diffraction, a new technique which uses very short wave-length X-rays to resolve molecular structure, rather in the manner that a microscope enables one to resolve the structure of cells. Rosalind Franklin and Maurice Wilkins at King's College, London, were already using X-ray diffraction to look at DNA. Another clue came from Chargaff, who had separated and measured the amount of the four nucleic acid bases in DNA from various sources. He found that the amount of guanine always equalled that of cytosine, and the amount of adenine always equalled that of thymine, in other words, G = C, and A = T. By piecing together this information with the X-ray pictures, Watson and Crick built the first model of the DNA double helix. Essentially, each molecule consists of two polymeric chains wrapped around each other by pairing their bases: G to C and A to T. Thus, the sequence of bases in one strand will be *complementary* to that in the other strand, rather like the positive and negative of a film. Based on that, a faithful copy of the complementary strand can always be obtained from one of the strands, and that was the basis of accurate reproduction of genetic information.

But how was the sequence of bases in DNA related to the sequence of amino acids in proteins? At around the same time, Claude Shannon had produced a very influential 'theory of information' in connection with the transmission of messages by telegraph. The idea of a 'genetic code' naturally suggested itself. The proteins can be seen as the final messages that are transmitted to the cell. Each amino acid is like a word in the message, and words must be specified by an alphabet. We know that there are only four letters in the genetic alphabet (the English language has 26 letters in its alphabet), these being the four bases. Thus, the minimum number of letters needed to code for 20 different words is 3, which gives 4^3 = 64 different codons.

From that original deduction, Francis Crick and Sidney Brenner set about analysing mutants of bacterial viruses in which different numbers of base pairs were deleted, in order to prove that it was indeed a triplet code. Successive triplets of bases on one of the DNA strands was read as an amino acid in a definite reading frame with no overlapping.

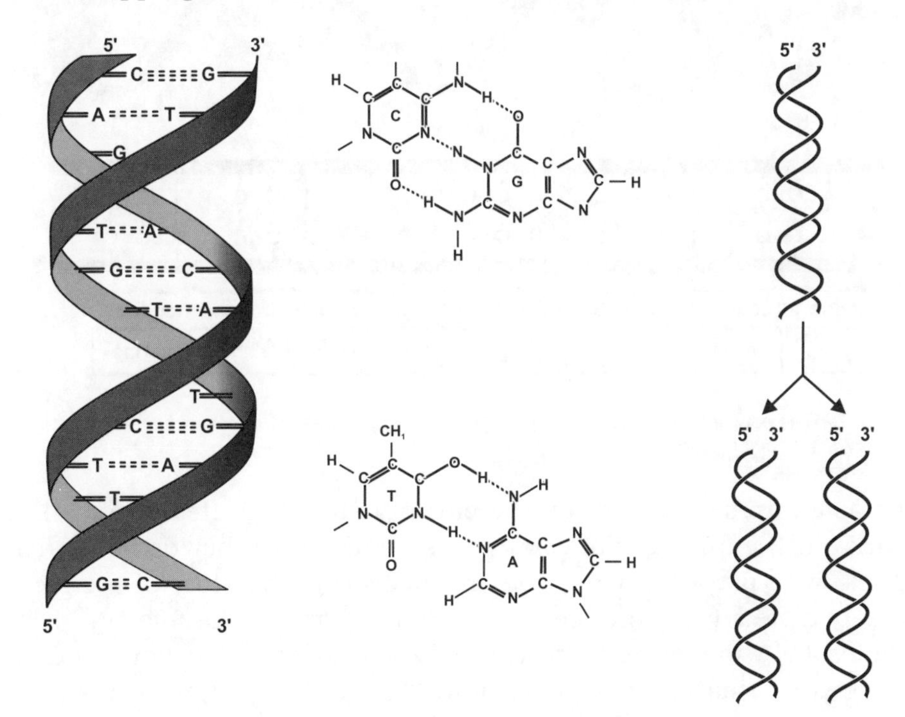

Figure 7.3 The DNA double helix and how it can be faithfully copied.

But which triplets specified which amino acid? That was the problem of 'codon assignment'. Before it could be done, Francis Jacob and Jacques Monod showed that the code in DNA is not translated directly, but goes through an intermediary or 'messenger RNA', transcribed from the gene, whose sequence is complementary to the DNA sequence of the gene. The messenger RNA is then transported outside the nucleus to the cytoplasm of the cell, where it is translated into protein. Marshall Nirenberg, Heinrich Matthaei and Severo Ochoa set to work on the genetic code. By adding synthetic messenger RNA of known base sequences to a cell-free system for making proteins in a test-tube, isolating and analysing the product, they were eventually able to make the complete assignment of codons by about 1966. Apart from codons for the amino acids, there are also codons for stopping and starting. As there are 64 possible triplets, some amino acids have more than one codon.

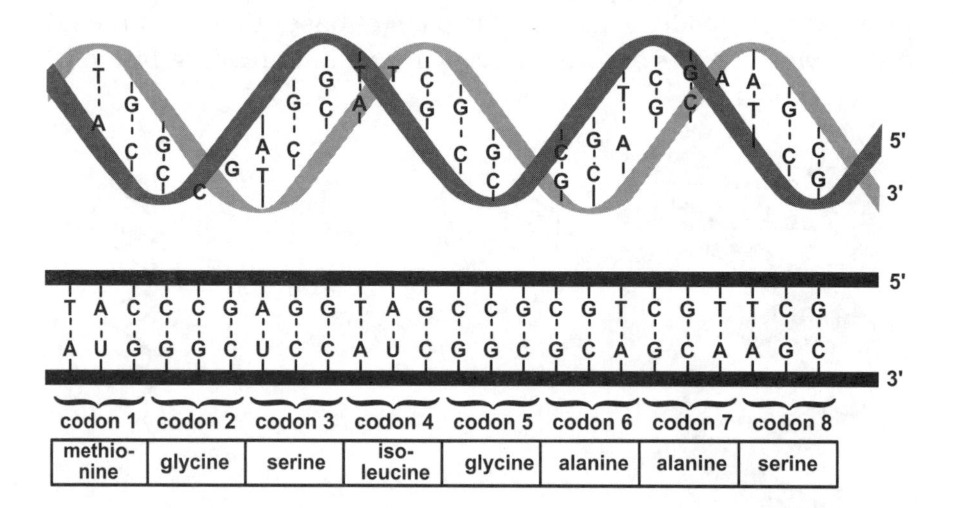

Figure 7.4 The flow of information from DNA to RNA to protein in accordance with the central dogma of molecular biology.

It was an immense effort on the part of many dedicated scientists to work out the full story of what genes do and how they do it. Francis Crick, who played a major role in uncovering the story of the gene, encapsulated it in the 'central dogma' of molecular biology: DNA makes RNA makes protein in a one-way information flow, and no reverse information flow is possible. The central dogma seems to be a direct vindication of Weismann's theory of the germplasm, which has since come to be known as 'Weismann's barrier'. This barrier is supposed to strictly forbid environmental influences, or any experience in the lifetime of the organism to directly (i.e. predictably) affect its genes. The entire process from DNA to proteins is depicted in Figure 7.4.

Many of the chief players in the story of the gene were physicists who made a conscious decision to devote their time and effort to work on the science of life, some because they had become distressed and disillusioned with the atomic bomb which became a 'science of death', a few because they were genuinely inspired by quantum physicist Schrödinger's little book, *What is Life?* published in 1944.[3] There is no doubt that the application of physical techniques such as X-ray mutagenesis, X-ray diffraction and physical concepts on information theory were responsible for the rapid advances. However, the physicists also reinforced the reductionist view that the story of life is just the story of the gene. This attitude has dominated biology to the present day.

The 'Selfish Gene' and the Vanishing Organism

In the new orthodoxy which has reigned supreme over the 20 years following the discovery of the double helix, the organism has become totally eclipsed from view. It tends to be regarded as no more than a collection of genes, and its development as the unfolding of a 'genetic programme' encoded in the genome. The reasoning goes that random mutations give rise to mutant characters and natural selection allows the fittest mutants to survive and reproduce. Environmental changes give new selective forces and evolution is thereby guaranteed. Thus, the organism is projected as a totally passive object, buffeted by selective forces in the environment over which it has no control. Its life experience is equally negated, as only the genes it carries are of any evolutionary consequence.

This ideology, which has underpinned the establishment of 'gene banks', supposes that so long as the genes or germplasm are preserved, the real organisms can go extinct. The recent rush to collect blood samples from indigenous tribes, in the disreputable Human Diversity Project, was inspired by the same mentality that decrees that one must preserve those valuable genetic resources as immortalised cell lines in the laboratory, before the indigenous tribes become extinct. This is something which many geneticists regard to be regrettable, but inevitable. Have they ever asked themselves if they've got their priorities right when they put the genes before the human beings to whom the genes belong? Should they not devote at least as much effort to helping those same indigenous tribes from extinction as they do to cheating them of their genes?

Richard Dawkins[4] has pushed the reductionist trend of neo-Darwinism to its logical conclusion in proposing that organisms are nothing but automatons controlled by 'selfish genes' whose only imperative is to replicate at the expense of other selfish genes. E.O. Wilson extended neo-Darwinian theory to animal and human societies to define the new discipline of sociobiology, which purports to explain all of social behaviour in animals, from ants to human beings, as the result of genes that have been selected in evolution. This has produced a veritable industry for many third-rate scientists with limited imagination who can think of nothing better to do than dream up selective advantages for putative characters controlled by putative genes, thereby becoming an instant success with their professors as well as the darlings of the equally simple-minded science journalists writing for the popular media.

In the opening pages of his book, E. O. Wilson gives the game away by posing the 'fundamental' and paradoxical (that is, paradoxical *within* neo-Darwinism) question of sociobiology: How could altruistic behaviour evolve, given that genes, and the behaviour they control, are fundamentally selfish? The paradox disappears, of course, when one rejects the assumption that selfishness or competitiveness is fundamental to the living world. Animals engage in competitive or aggressive acts, but that does not mean there are inherent qualities of competitiveness or aggressiveness which can account for those acts. Furthermore, examples of cooperation among animals far outstrip those of competition. Kropotkin,[6] social anarchist and Russian prince, argued that cooperation, or mutual aid, was much more important than competition in the evolution of animals and of our own species. He gave abundant evidence of the natural sociality of all animals which is independent of genetic relatedness. Animals, including human beings, simply enjoy society for its own sake. Thus, one could easily invert Wilson's question and ask, why do animals compete, given their natural sociality?

This highlights the socio-political underpinnings of all scientific theories - yes, my own being no exception. Perhaps the advantage of being a foreigner, a culturally disposessed one at that, and having to adopt a foreign culture as my own, is that I find I cannot take anything for granted. Maybe that makes it easier to transcend the socio-political underpinnings of the domiant culture. (At least, that is my story!). Darwin's theory is all of a piece with Victorian English society's preoccupation with competition and the free market, with capitalist and imperialist exploitation. Unfortunately, this same ideology is very much alive today, and is being played out in the World Trade Organisation negotiations currently going on.

It is no accident that a culture bent on promoting capitalism and free enterprise should be obsessed with *things* rather than processes. The notions of 'gene banks' and 'genetic resources' make it plain that life, the process of being alive, as well as real organisms and diverse ecological communities, are all negated in favour of genes which can be grasped hold of, possessed, preserved and exploited as commodities.

Notes

1. Morgan, 1916, pp. 187-190.
2. More details on the experiments performed, from the identification of DNA as the genetic material to the cracking of the genetic code, are to be found in Ho, 1976.
3. Schrödinger, 1944.
4. Dawkins, 1976.
5. Wilson, 1975.
6. See Kropotkin, 1914, also Ho, 1996c for a more extended discussion of the point raised here.

Chapter 8

The Fluid and Adaptable Genome

Would anyone think of investing in genetic engineering biotechnology if they knew how fluid and adaptable genes and genomes are? The notion of an isolatable, constant gene that can be patented as an invention for all the marvellous things it can do is the greatest reductionist myth ever perpetrated. Genes and genomes need to be fluid in order to maintain stability. That is the essence of organic stability, as opposed to mechanical stability.

The End of Mechanistic Biology

The new genetics spells the end of mechanistic biology, which has dominated the world for at least a hundred years. In order to appreciate the profound conceptual change involved, let me remind readers who are old enough of the textbook genetics that we were taught right up to the early 1970s and possibly beyond, which came straight out of the central dogma of molecular biology that reigned supreme from the 1950s to the 1970s (see Chapter 7).

The Central Dogma of Molecular Biology of the 1950s to 1970s[1]

1. *DNA (and in some viruses, RNA) is the genetic material.*
2. *Genetic information flows from DNA to RNA to protein via a triplet genetic code.*
3. *The base sequence of the RNA transcribed, and subsequently translated into a polypeptide, is a faithful complementary copy of the DNA encoding the polypeptide.*
4. *One polypeptide is specified by one gene, which is a continuous sequence of DNA.*
5. *The sequence of bases in the gene corresponds exactly to the sequence of amino acids in the polypeptide it encodes.*
6. *The genetic code is universal.*
7. *The sequence of base triplets in a gene is read in one direction, without overlap, and only in one correct reading frame.*

8. *The DNA of most cells remains constant during development, only the genes expressed differ between different types of cells.*
9. *Environmentally induced modifications in the characters of somatic cells do not affect the DNA and cannot be inherited.*

Thus, the genes determine proteins in a completely mechanical sequence which is well illustrated by the sort of diagram given in Figure 7.4 in the previous chapter. There are a few complexities - for instance, the fact that not all genes are expressed in all cells, so there must be other genes which control the genes being turned on and off. But what controls the genes that turn other genes on an off? This mechanical hierarchy of genes controlling genes controlling other genes etc. soon gets into an infinite regress that fails to explain how living organisms can become organised. But these are essentially the sort of explanations we are offered to this day.

Beginning in the early 1970s, all but the first assumption of the central dogma have become violated. The first assumption remains true only by definition, as it is well-known that certain cellular states, or gene-expression states, are heritable, quite independently of changes in DNA or RNA (see later). The initial crack appeared before rDNA research really got underway. It was the discovery, by US geneticists Howard Temin amd David Baltimore, of a viral enzyme, *reverse transcriptase*, that does the reverse of transcription - that is, it makes a copy of complementary DNA (cDNA) from an RNA sequence. The kind of viruses possessing reverse transcriptase are *retroviruses*, implicated in AIDS and many forms of cancer, which have RNA as their genomes.

Overlapping genes were first discovered in the bacteriophage ϕX174 (a virus of bacteria) when its complete base sequence was worked out by Sanger in 1977. The phage possesses at least one gene that codes for two separate polypeptides. The read-out for the second polypeptide starts one base out of phase in the middle of the gene and continues on to the end, so the second message gives a totally different polypeptide. Overlapping genes have now been found in many other viruses.

Then, it was discovered that mitochondria DNA (DNA inside the cellular organelles mitochondria where organic substrates derived from food are oxidised to provide energy for all kinds of vital activities) uses a genetic code that differs from the 'universal' code in several respects. Subsequently, the ciliated protozoa (single celled animals) such as *Paramecium* and *Tetrahymena* were also found to use yet another code. So the genetic code is by no means universal.

Surprising though those findings were, they could easily be dismissed as exceptions which prove the rules. By far the most significant picture to emerge from the violations of the central dogma of molecular biology is that of the dynamism and flexibility of the genome in both its organisation and function. This is in striking contrast to the relatively static and mechanical conception that previously held sway.

The Case of the Vanishing Gene

In the mindset of genetic engineers, the 'gene' is still very much the same as the one that molecular biology of the 1950s to 1970s established: it is a continuous stretch of DNA with a specific base sequence (remaining unchanged except for extremely rare random mutations) located in a definite position within the constant genome of the cell that determines the amino acid sequence of a corresponding polypeptide. The amino acid sequence of a polypeptide, in turn, determines its function in the organism. This notion of an isolatable gene specifying a function independent of the cellular and environmental context of the organism, is also the one which validates the patenting of genes. The first reductionist fallacy in the patenting of genes is that DNA, by itself, can specify nothing at all, as DNA depends, for its replication, on the entire cell. What the findings of the new genetics also show is that the gene itself has no well-defined continuity or boundaries, the expression of each gene being ultimately dependent on, and entangled with, every other gene in the genome. There is certainly no one-way information flow proceeding from DNA to RNA to protein and the rest of the organism, as projected by the central dogma. Instead, gene expression is subject to instructions, modifications and adjustments, according to the environmental, physiological and cellular contexts. Moreover, the base sequence of DNA in genes and genomes is subject to small and large changes in the course of normal development and as the result of environmental perturbations.

That whole period from around the mid-1970s to the mid-1980s was one of the most exciting and rewarding for biology. It was as though a hundred flowers were blossoming after the tyranny of the central dogma. Every week, it seemed, a new discovery was made that contradicted what had been accepted for decades previously. And it was a truly collective endeavour. Hundreds, if not thousands, of molecular geneticists in academic institutions all over Europe and the United States were involved, sharing information, exchanging researchers, collaborating and cooperating in any and every way possible. There were still 'heroes and heroines', but a typical paper in molecular genetics contained ten or more authors, so team work was taken for granted. There was a lot to be done and so many new things to be discovered that no one really worried about guarding any secrets. Although they probably had no time to reflect upon it then, those researchers were building a new genetics paradigm for the next millennium. The notion of an isolatable, constant gene that could be patented as an invention for all the marvellous things it could do had never crossed their minds. And, if it did, and if they had reflected on the implications of the totality of their discoveries, they would have recognised that notion for what it was: the greatest reductionist myth ever perpetrated, that flies in the face of all the scientific evidence.

Let me try to convey some of the excitement of that period, to include findings which continued to be made well into the 1990s.

The interrupted gene

The discovery of interrupted gene made headlines in the late 1970s, as it was totally unexpected. The discovery was made possible by *gene cloning*, the technique of making many copies of a gene, isolating the gene and identifying it. When a specific gene is isolated, its base sequence can be determined and compared to that of the mRNA transcribed from the gene, as well as the amino acid sequence of the polypeptide translated from the mRNA. By carrying out those procedures, researchers found that the gene corresponding to coding sequence in the genome is actually interrupted at intervals by long stretches of non-coding sequences. The coding regions came to be known as *exons* and the non-coding regions as *introns*. This structure is now found to be characteristic of most eukaryotic genes. The number and size of introns vary greatly, and they are often much longer than the coding sequences (you will come across some human genes in Chapter 13). At transcription, the complementary sequence of the entire gene is transferred into a precursor RNA or primary transcript, which is then further processed into the messenger RNA. Processing turns out to be *very* complicated and involves, among other things, splicing out the introns, so that the complete coding sequence can be translated into a continuous polypeptide chain. Thus, the gene sequence in the genome does not at all correspond to that predicted from the amino acid sequence of the polypeptide encoded.

Interrupted genes were a surprise, but it is really the discoveries of the many layers of complexities involved in gene expression that gives rise to a more profound change to the status of the gene. The gene responsible for making a single polypeptide is functionally, as well as structurally, ill-defined. It is de-localised throughout the genome, being entangled with all other equally ill-defined genes. As the expression of the gene - the synthesis of the polypeptide - is also sensitive to physiological and environmental conditions, the gene is ultimately de-localised over the entire organism in its ecological setting. I shall explain why that is so.

The de-localised, entangled gene

The DNA sequence by itself can do nothing, as it depends on enzymes and other proteins interacting with it to be replicated, and to be transcribed. Gene expression - the eventual appearance of the polypeptide encoded - turns out to be an extremely complicated process. It depends on special 'regulatory' DNA sequences which may be found in front of (5' to) or behind (3' to) the region of the (interrupted) gene, within the introns themselves, or sometimes very far away on the chromosome. These sequences interact with a host of regulatory proteins, or *transcription factors*, encoded by other genes scattered throughout the genome (on other chromosomes). Each of these genes will most likely possess a gene structure like the one they are regulating, and requiring other transcription factors for regulation. Each transcription factor recognises a special short sequence *motif* within the regulatory region(s). Transcription cannot start until the *transcription complex*, consisting of several transcription factors, are bound to one of the regulatory regions, the *promoter*, that marks the start-site of transcription.

Although some transcription factors are common to all cell types, and are essential for the transcription of many, if not all genes, other transcription factors are cell specific, or are activated by specific stimuli, such as heat shocks, specific metabolites or hormones, which are themselves subject to environmental modulation.

Another illustration of the de-localised, mutual entanglement of gene functions is in the gene-polypeptide relationships. According to the central dogma, one gene encodes one polypeptide. In reality, *all possible mappings* exist between genes and polypeptides: one to one, one to many, many to one, and many to many.[2] For example, alternative splicing of the primary transcript (to remove introns) gives rise to more than one protein from the same gene. This occurs in different tissues as well as in the same tissues. In some cases, a large polyprotein encoded by one gene is split, after translation, into two or more polypeptides. More surprisingly, many genes can be joined together by DNA rearrangement in the genome, which is then transcribed and translated to give one polypeptide. This occurs in the synthesis of the *immunoglobulins* or antibody proteins that bind specific foreign antigens, as the body mounts an immune response. By recombining different variants in multigene families (see below) of several different genes, a huge diversity of antibodies can be made, which are specific for binding each of the thousands of different foreign antigens that the organism is likely to come across.

Finally, many genes can encode the same or similar polypeptides in so-called *multigene families*. These are families of genes that exist in multiple copies (from several copies to many thousands or hundreds of thousands of copies) in the genome. *Simple* multigene families have identical or nearly identical sequences that are arranged in one or more tandem arrays (continuous head to tail repeats). These function simultaneously to make large amounts of single proteins, such as the *histones* that package DNA into chromosomes in eukaryotic cells, and ribosomal RNA, involved in translating mRNA into proteins. *Complex* multigene families, on the other hand, contain sequences that are not identical, though they are similar and serve related functions. Examples are the *haemoglobins*, the oxygen carrier proteins of the red blood cells, which exist in several embryonic, foetal and adult forms.

The gene that gets silenced

Although genes encode amino acid sequences of proteins, they may not become expressed, subject to 'instruction' from the cellular and physiological states. Approximately half of all genes in the eukaryotic genome have a large number of repeats of the dinucleotide, CG, at their front (5') ends. The cytosine residues often have a methyl group, $-CH_3$, added to them. It turns out that, when a lot of them are methylated, the gene is no longer transcribed. This regulatory mechanism operates not only on normal genes in the genome, but also on transgenes, much to the frustration of genetic engineers (see Chapters 9 and 10), and on integrated viral sequences. This mechanism can be triggered simply by introducing more copies of a gene that already exists in the genome, and is part of the defence mechanisms against foreign, unwanted DNA that all cells possess.

A major silencing or inactivation of genes has already been known for a long time, although it is still not yet fully understood. It involves one or other of the two X chromosomes carried by the female of many species, including human beings, where the males have only one X chromosome and a very much smaller Y chromosome, with only a very few genes. Inactivation of one of the two X chromosomes in the female starts at one end and spreads eventually to the entire chromosome, which ends up looking like a contracted round blob in the nucleus. The genes on the inactive X are all heavily methylated, except for a few that are still active. It seems entirely a matter of chance which X chromosome in a cell becomes inactivated, so heterozygotes will end up being a somatic mosaic - a mixture of different cells. One of the most common mosaics is the tortoiseshell cat, a female heterozygous for an X-linked gene, which gives either orange or black coat colour. The result of random X chromosome inactivation is a mosaic of orange and black patches which is most attractive.

What is interesting about DNA methylation is that it is inherited, so that, when the DNA is replicated, the bases marked by methylation in the old strand are also methylated in the new strand. This so-called 'epigenetic' inheritance is at least partly responsible for the stability of the differentiated states of cells. When certain genes lose methylation patterns as a result of changes in cellular states, this is also passed on to the daughter cells. Thus, both the acquisition and loss of methylation are among the many examples of the inheritance of acquired characters,[3] which orthodox opinion has been at pains to deny. These and other cases of changes in gene expression that are inherited do *not* require changes in DNA sequences.

The gene that gets edited

Since the 1970s, geneticists have uncovered more and more complications in the 'processing' of the primary RNA transcript into messenger RNA that gets translated into protein. The most surprising of all is *RNA editing*, in which the base sequence of the RNA transcript is actually changed by the addition of bases to the RNA molecule or by the chemical transformation of one base to another. The process was originally discovered in 1986 in the mitochondrial transcripts of the trypanosome, but it has since been found in organisms as diverse as mammals, amphibians, plants, protozoa and viruses, involving not only mitochondrial and chloroplast genes,[4] but also nuclear genes.[5] It is likely to be a very common process in gene expression. RNA editing does not occur at random, but whenever it occurs, it is probably *essential* for proper functioning of the organism. For example, in wheat mitochondria, the transcript of a gene, *atp9*, is edited by removing an amino group, -NH_2, from cytosine, C, turning it into Uracil, U. The edited mRNA gives rise to a functioning protein with an amino acid sequence that differs from that encoded in the gene. Unedited proteins introduced in transgenic plants resulted in male sterility.[6] RNA editing depends on RNA 'editases', enzymes that exist in different forms specific for different editing jobs.[7] Moreover, multi-protein complexes are involved in editing, as in transcription.[8]

Thus, the gene may not even determine the amino acid sequence of the polypeptide it is supposed to encode. Instead, the precise sequence of amino acids depends on influences *from* the context - the cellular and physiological state - propagating backwards to the post-transcriptional levels, just as similar backward influences are propagated to the level of transcription through transcription factors and DNA methylation to determine which genes are transcribed and which genes are silenced. The causal loop for gene expression is circular and multidimensional. There is no simple, linear, unidirectional instruction proceeding from the gene to RNA to protein. We shall have the opportunity, later, to examine many 'reverse information flows' that proceed from the environment backwards to *alter* genes and genomes.

What is becoming clear is that the mechanical concept of an isolatable sequence of DNA corresponding to a gene does not accord with the organic reality of the dynamic, de-localised entanglement of gene function. The transition between the molecular genetic determinism of the central dogma and the new genetics is reminiscent of the transition between the separate, mechanical objects of the Newtonian universe and the de-localised, mutually entangled entities of quantum reality.[9] The new-found dynamism in gene function is fully matched by the fluidity of genes and genomes.

The Fluid Genome[10]

"The application of new molecular techniques reveals that, beneath the level of the chromosome, the genome is a continuously changing population of sequences. Mobility, amplification, deletion, inversion, exchange and conversion of sequences create this unexpected fluidity on both an evolutionary and developmental time-scale."[11] This quotation is from an historic publication, *Genome Evolution*. By the use of this title, the co-editors, Gabriel Dover and Richard Flavell, have, in effect, defined a completely new subject area.

Genome organisation is infinitely variable

The genomes of *eukaryotes* ('higher' organisms whose genomes are enclosed in a membrane-bound *nucleus* in the cell, as opposed to *prokaryotes*, such as bacteria, which do not have a nucleus, and whose genomes exist free in the cytoplasm of the cell) are very big and messy. They are also infinitely variable, as geneticists came to realize when they had a chance to dissect eukaryotic genomes with recombinant DNA techniques.

First of all, there is far too much DNA than is required to code for all the proteins, and to supply all the signals necessary for gene transcription. The overwhelming proportion of the DNA - perhaps up to 99% in some genomes - appears to have no known function. It has been described as 'junk DNA' or 'selfish DNA' - selfish because it serves no purpose except to get itself replicated along with the rest of the genome. Secondly, most of the DNA consists of repeated sequences. Repeats vary in number from less than ten to hundreds of thousands, or several millions. The length of the sequence repeated varies from two or three base pairs to thousands or hundreds of thousands of base pairs. These repetitive sequences may be clustered near the ends of chromosomes or centromeres, in other parts, or dispersed throughout all parts of all the chromosomes. The number of repeats and, sometimes, also their location, differ between individual genomes belonging to the same species or populations. Repetitive sequences are making DNA sequencing of the human genome extremely difficult, even apart from the obvious question of whose genome it is that is being sequenced, as individual genomes are unique in their DNA sequences, and in their organisation. (You will see in Chapter 12 how variable individual genes coding for proteins may be, and how the enormous variation in repeated DNA sequences in the human genome can be used to locate genes and provide identification of individuals for forensic purposes.) Although some of the repeated sequences are members of multigene families that code for functional proteins, the vast majority of them do not have any known function.

Some repeats are mobile genetic elements, also called *transposons* or 'jumping genes' - sequence elements that encode genes for enzymes that can excise and re-insert the elements in different locations in the genome, in the course of which they may make further copies of themselves. Transposons were first discovered by geneticist Barbara McClintock more than 50 years ago. For this, she was awarded the Nobel prize, very belatedly in 1983, and has ever since been elevated to the status of folk hero. She was studying a number of very unstable genes in maize that mutated spontaneously at high frequencies. This was due to transposons that jump in and out of genes, disrupting their function. Even when the transposon jumps out again, function may not be restored, as transposons usually leave behind their 'footprint', which is a duplication of short sequences flanking the site of insertion.[12] (Transgenic maize is probably expecially hazardous in that regard, as there are already plenty of transpones providing helper functions to mobilise transgenes and marker genes.)

One class of transposons are *retrotransposons*, which depend on a reverse transcription step to move and to duplicate. These are very similar to retroviruses, from which they may have originated (see Chapter 13).[13] A host of other retroviral-like relict sequences, that have lost their ability to mobilise independently, are also present in the genome. These relict sequences do not necessarily cease to mobilise and duplicate themselves, for they can be helped by other elements. That is why the so-called 'crippled' vectors of genetic engineering biotechnology are so dangerous (see Chapters 10 and 13). In all cases, reverse transcription has made a cDNA copy from the corresponding RNA, which is then inserted into the genome.

Reverse transcripts turn out to be very common in eukaryotic genomes. Up to 20% of some genomes may consist of reverse transcripts. One class of such reverse transcripts, called *Alu* sequences (so named because they are cut by one of the many DNA cutting enzymes, or *restriction enzymes*, called Alu) are the most abundant middle repetitive DNA sequences in the human and rodent genomes. In the human genome, it is a 300bp sequence repeated some half a million times, the repeats being widely dispersed in the genome. It is homologous to parts of an RNA molecule in the cytoplasm, referred to as the 7SL RNA, which makes up the signal recognition particle (SRP) - part of the molecular machinery required for moving proteins across intracellular membranes after they have been translated. The 7SL RNA consists, intriguingly, of an *Alu* sequence into which a sequence specific to 7SL RNA has been inserted.[14]

Finally, the genome also contains many 'pseudogenes' - nonfunctional coding sequences that have been reverse transcribed from their mRNAs and re-inserted into the genome. Reverse transcription constitutes a potential route of reverse information flow (see below).

Genome dynamics - DNA turnover[15]

The reason genomes are so big and messy is because they are continuously changing due to many processes, operating constantly on developmental and evolutionary time-scales. These processes destabilise genes and genomes, move genes around, mutate, rearrange, recombine, replicate sequences, delete or insert sequences, and even exchange and convert sequences. Sequences in the genome can be amplified (or contracted) thousands and hundreds of thousands of times as part of normal development, or as the result of environmental challenges. They can undergo large reorganisations or rearrangements to form new chromosomes. Genomes can duplicate wholesale, subsequent to the formation of hybrids between species whose chromosomes are not sufficiently similar for them to pair up properly in the special cell divisions (*meiosis*) that form germ cells. In such a case, doubling the hybrid genome wholesale allows every chromosome to pair up with a partner identical to itself. This process - *polyploidization* - is very common in the evolution of natural species. That is why closely-related species may have genomes of vastly different sizes. The biggest genomes in the world belong to single-celled ciliate protozoa, which tend to amplify their entire genomes as part of normal development.

The fluidity and dynamic nature of the genome so impressed molecular geneticist Richard Flavell that he envisaged cycles of 'DNA turnover' in the genomes of all species, involving mutations, rearrangements, translocations, amplifications and deletions, providing major sources of variation for the evolution of new species.

Jumping genes

Transposable elements are responsible for much of the fluidity of genomes. Transposition not only leads to changes in the position of the transposable elements themselves but, in the case of replicative transposons, the process can spread copies of the transposons around the same genome or to other genomes by infection. One transposon, the *P* element, has spread to all *Drosophila* species within a period of 50 years.[16] *Drosophila* strains in the laboratory, which were established before the spread of *P* elements, were free from the element. A tiny mite, parasitic on many *Drosophila* species, is thought to have been responsible for spreading the *P* element across species barriers.[17] You will come across other examples of transposable elements in Chapters 10 and 13. Transposable elements are responsible for many 'spontaneous' mutations in maize, as mentioned above, in *Drosophila* and also in human beings. Transpositions lead to sequence duplications, deletions and chromosomal rearrangements. The genetic upheavals caused by transpositions can be quite considerable.

Transposable elements in the genome do not always move. Transposition appears to be regulated by cellular functions, so transposons may change from an active to an inactive state. Inactive transposons are correlated with an increased level of methylation[18] which is strongly implicated in 'gene silencing' - the failure of gene expression (see above). The frequency of transposition is greatly increased as the result of environmental stress, in both maize[19] and *Drosophila*,[20] however. While some geneticists see this to be a potentially adaptive function, generating variants which may enable the organism to overcome the environmental challenge, the overall result of stress-induced transposition is to increase the rates at which genes become inactivated as the transposons insert randomly into them. This has implications for human health (see Chapter 14).

Amplifying and contracting genes

Gene amplifications and contractions can occur as part of normal development. For example, it has been known since the 1960s that the genes encoding ribosomal RNA, which are necessary for translating mRNA into proteins, undergo waves of amplifications during maturation of the germ cell. This happens again in the fertilised egg of *Xenopus* (the African clawed-toad, which is really a frog), so that eventually, more than 70% of the nuclear DNA in the egg codes for rRNA. Ribosomal RNAs already exist in the genome as multi-gene families. Amplification does not involve the entire cluster of repeats. Instead, there is often a predominant class which is amplified in the cluster.

Gene amplifications and contractions occur in many plants which switch from juvenile to adult phases. This involve changes in the repetitive sequences of all the cells, presumably in the growing meristem of the shoot.

The potential for gene amplification and contraction is actually very widespread. It happens readily in mammalian cells in the course of drug treatment, as in chemotherapy against cancer, in which the cells develop resistance to the drugs One of the best-studied examples is methotrexate resistance.[21] Methotrexate inhibits the activity of the enzyme dihydrofolate reductase, which is necessary for making DNA, so that cells unable to make DNA will no longer multiply. Resistance to methotrexate can be due to increased activities of membrane proteins which pump drugs out of the cell, or it can involve mutational changes in the dihydrofolate reductase enzyme, so that the new enzyme is no longer inhibited by the drug. A third mechanism is the over-production of the normal enzyme, so that there is an excess of the enzyme in the cell which overcomes the inhibition by the drug. Over-production of the enzyme is accomplished by gene amplification. Whole segments of the chromosome containing the gene is amplified, the amplified unit being at least ten times as long as the gene itself. Amplifications are accompanied by gross changes in chromosome morphology, chromosomal rearrangements, and by the acquisition of extra chromosomes exhibiting a wide range of abnormal morphologies.

These changes in the genome are repeatably generated by particular drugs in particular cell lines. *They have nothing to do with selection of random mutations*, but are physiological responses shared by all the cells in the population. Similar genetic and genomic changes are induced in insects exposed to insecticides and plants exposed to herbicides (see next chapter). They are part and parcel of the spectrum of physiological responses common to all individuals in a population. In Chapter 11 you will find further examples in the origin of antibiotic resistance in bacteria. The extent to which the genomes of both somatic and germ cells can change in response to the environment should make geneticists extremely cautious about 'cloning' from the cells of adult animals (see Chapter

10), as well as making inferences about the primary causes of diseases such as cancer (see Chapter 14).

Gene amplifications are known to be involved in cancer itself (see Chapter 13). All gene amplifications share certain common features. The amplified sequence is specific for each developmental or environmental stimuli; and the amplified DNA contains a large proportion of extraneous sequences other than the recognised gene sequence, which presumably also contains the regulatory signals for transcribing the gene.

Environment changing genes[22]

One interesting class of environmentally-induced DNA changes occurs in association with a whole spectrum of heritable modifications subsequent to treatment with various mixtures of fertilisers in flax and other plants. As is well-known, plants do not have separation between somatic and germ cells, for every somatic cell is capable of developing into germ cells, so somatic modifications will be inherited in subsequent generations.

This phenomenon was first discovered in the 1950s, but was not analysed by molecular genetics techniques until the 1980s by Chris Cullis,at the John Innes Institute in Britain. Chris found that a 1.5% solution of the fertiliser, ammonium sulphate, added to compost at the time of sowing, gave a large type, L, whereas a 1.5% solution of a triple-superphosphate with low pH compost, gave a small type, S. The L and S 'genotrophs', as they are called, differ in many characteristics from each other and from the parental type from which they were derived. Differences found include size, hairiness, distinct types of enzymes expressed, as well as genome size and the copy number of repeated ribosomal RNA genes. The changes are stably inherited in subsequent generations in the absence of the inducing environment, under the usual conditions of growth. However, the stability is not absolute, and further changes can be induced by other circumstances.

Careful analysis of seedlings treated with fertilisers at different stages of growth showed that the DNA changes take place in *all* the cells of a meristem (the growing zone of a plant) *simultaneously*. Thereafter, the growth characteristics of the meristem tend to improve, suggesting that at least some of the changes are adaptive. These environmentally-induced DNA changes are specific for different environments, and can be repeatably generated. They are, therefore, *not* random changes. Similar environmentally-induced DNA changes have since been documented in other plants, such as maize, pea and broad bean.

Gene conversion and concerted evolution[23]

A puzzling phenomenon has emerged as multigene families from different species have been analysed. It has been found that the members of multigene families within a species are much more homogeneous than expected if each member gene accumulates mutations independently of all the rest. Furthermore, each species is homogeneous for variants that are diagnostic of the species. In other words, variation is accumulating between different species but, within a species, all members of each multigene family tend towards uniformity. Actually, there are degrees of uniformity, as each multigene family can consist of sub-families, the members of which are more similar to one another than they are to members of a different sub-family. But the puzzle remains the same. What is responsible for this 'concerted evolution' of sequences, many of them dispersed throughout the genome? Gabriel Dover achieved a degree of notoriety among the more orthodox community by drawing attention to this phenomenon very early on, calling it by the colourful term, 'molecular drive', on the grounds that it drives evolution much more substantially and rapidly than natural selection.

There are several mechanisms that can result in uniformity of sequences. The first is gene amplification (see above) in which an enormous number of tandem repeats are generated all at once. Another is unequal exchange between homologous chromosomes which occurs during chromosome pairing in the formation of germ cells, so that one chromosome acquires more copies and the other less. However, these mechanisms cannot explain the homogeneity of dispersed copies, and even tandemly-repeated copies will tend to accumulate mutations independently. A third mechanism is through replicative transposition, but, again, the replicated copies will still tend to diverge after the event. So the evidence points to some process(es) of gene conversion - changing DNA sequence from one to another. In the case of multi-gene families, it appears that all the family members tend to be converted to a uniform sequence, so that those sequences that have diverged are simply eliminated.

When sequences belonging to the 100 or so ribosomal RNA genes from *Neurospora* (the bread mould) were analysed, they were found to belong to 4 sub-families. Within each sub-family, the coding regions differed in sequence by less than 0.2%, whereas the flanking 'spacer' regions differed by 3.5 to 7%. Differences between the sequence of each sub-family were such that they preserved the same secondary structure of the ribosomal RNA necessary for its function, suggesting that gene conversion may depend on gene function. One way in which this may work is via reverse transcription, which appears to be very active in eukaryotic genomes, for which no other obvious function exists. Gene product ribosomal RNAs that are efficiently transcribed and stable escape destruction by ribonucleases that break down RNAs in all cells. This means that only genes that function well may be reverse transcribed into complementary DNA (cDNA) which is inserted into the genome, perhaps in place of those sequences that do not work well. A similar mechanism has been postulated for transfer RNA (tRNA) genes in yeast. The tRNAs are yet other genes required in the complicated process of translating mRNA into protein.

Gene conversion is an example of a mechanism that operates not only at the level of the whole organism, but at the level of the entire species. It is also a clear instance where the fluidity of the genome is involved in actively maintaining the stability of genes and genome. This is what organic, as opposed to mechanical, stability is all about. Genes and genomes need to be fluid in order to maintain stability (see Chapter 14).

Wandering genes

As mentioned above, transposons can travel between species that do not interbreed. The extent to which horizontal gene transfer has contributed to genome evolution is a subject of dispute among geneticists. However, the full scope of actual and potential horizontal gene transfer has come to light within the past 3-4 years. There have been 100 or more papers published in mainstream journals since 1993, all but two giving direct or indirect evidence of horizontal gene transfers. Transfers occur between very different bacteria, between fungi, between bacteria and protozoa, between bacteria and higher plants and animals, between fungi and plants, between insects. A transposon, called *mariner*, first discovered in *Drosophila*, has recently been found to have jumped into the genomes of primates including humans, where it causes a neurological wasting disease (see Chapter 13).

The current state of our understanding is presented in Fig. 8.1, where the arrows indicate transfers for which direct or circumstantial evidence already exist. Genetic engineering biotechnology poses special, unique dangers in that it greatly facilitates horizontal gene transfer through vectors that are designed to break down species barriers and to overcome cellular mechanisms that break down or inactivate foreign DNA. Horizontal gene transfer will be dealt with in greater detail in Chapters 9 and 10.

The Inheritance of Acquired Characters

The issue of the inheritance of acquired characters has been hotly debated for almost a century. The reality, which would take whole books to describe properly, is that there is no longer any doubt that acquired characters are inherited, and in many different forms. A number of them have already been identified In this chapter, I shall briefly review these below.

Epigenetic inheritance[24]

Cellular gene expression states, such as the pattern of DNA methylation, are inherited in somatic cells. In species which do not have segregation of germ cells and somatic cells, which include plants, fungi, bacteria, unicellular animals and many multicellular animals, these patterns will be inherited in subsequent generations. In vertebrates, a re-methylation of genes that are tissue-specific in expression occurs as part of the 'reprogramming' of the genome. But sometimes genes escape this reprogramming, leading to germline inheritance of acquired gene expression states.

There are many other forms of such epigenetic inheritance that do not necessarily involve changes in DNA base sequences, among them, the ill-defined cytoplasmic states that give rise to cytoplasmic inheritance which is not directly associated with DNA or RNA. They are known as *dauer-modifications* (lasting modifications), and tend to diminish in successive generations. In my laboratory, we documented a case of such cytoplasmic inheritance in the fruitfly some years ago.[25] It may be similar to the changes that occur in cells exposed to carcinogens, X-rays and metabolic stress. These epigenetic changes, rather than mutations in putative cancer genes, have been proposed to be the primary cause of cancer (see Chapter 14).

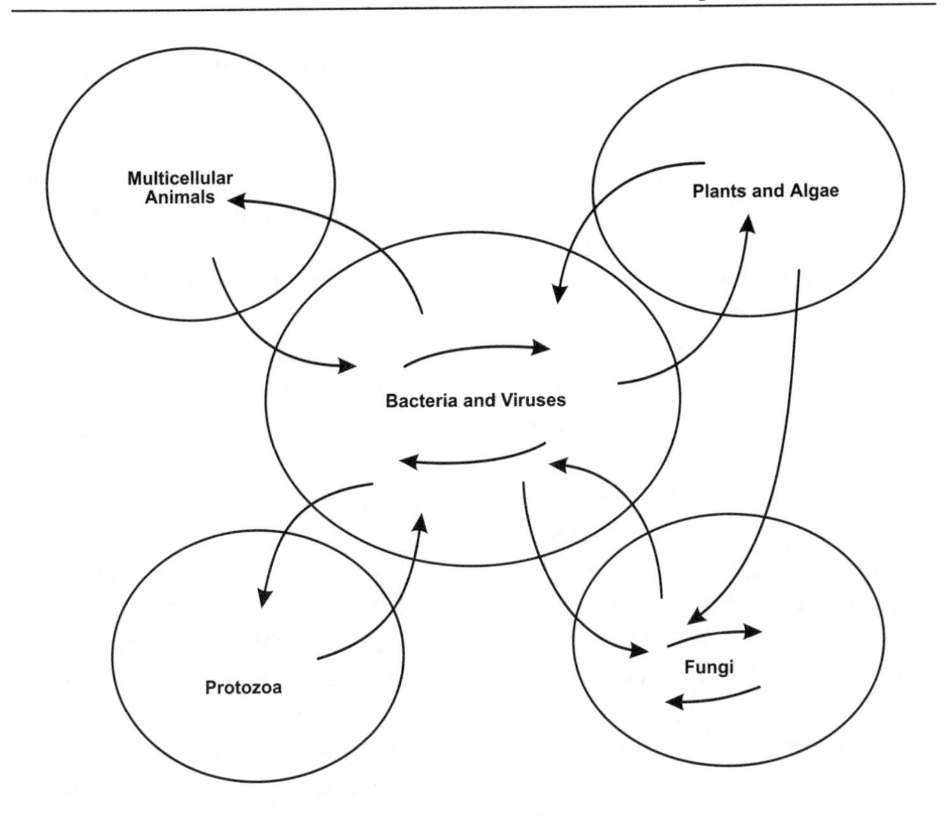

Figure 8.1 Horizontal gene transfer links the whole biosphere

Inheritance of induced changes in genomic DNA

As described above, the inheritance of induced changes in genomic DNA occurs in cells and organisms exposed to various toxic substances including cytotoxic drugs, insecticides and herbicides (see also Chapter 9). They also occur in plants exposed to various fertiliser treatments. As plants and the majority of animal phyla do not have segregated germ cells and somatic cells, these modifications will be inherited in subsequent generations. Even in animals with apparently segregated germ cells, the germ cells may also respond directly to the same stimuli, or have communication channels with somatic cells. One such channel is via reverse transcription.

Feedback from somatic cells to germcells

Again, you have already seen this illustrated in the intriguing phenomenon of gene conversion. There seems to be a cellular mechanism that effectively reports back to the germline, to change the genes or to stabilise them according to the experiences of somatic cells. Reverse transcription, as you have seen, is extremely common in the genome. This has led to many speculations that it may be a normal physiological regulatory mechanism[26] that maintains the stability of organisms on the one hand, and prepares them to deal with environmental changes on the other.

Reversed transcripts are abundant in the genome. Many are cDNA copies of the mRNA of cellular genes, and are largely non-functional. But functional 'pseudogenes' have been identified in the genome. For example, one of the rat insulin genes was shown to be part of a functional retrotransposon.[27] Moreover, there is evidence suggesting that a certain family of repeated DNA sequences in primates, the L1, may be derived from a sequence encoding a protein which is a reverse transcriptase.[28]

Immunologist Ted Steele proposed back in 1979 that acquired immuno-logical tolerance to foreign antigens is inherited *via* the male line, the implication being that it involves specific changes in germline DNA. That was a bombshell in those days. With characteristic proselytising zeal, Ted did not just publish the result of his experiments in the top journal, *Nature*, after wrangling with the then editor, John Maddox, but went on a world lecturing tour to publicise the results as widely as possible. Some of us cheered for him from the sidelines, but the orthodox community felt so threatened that they did their best to discredit him. However, he had been right all along, only the details needed filling in. His laboratory has now obtained evidence suggesting that somatically mutated immunoglobulin genes produced by B cells (white blood cells involved in the immune response) are 'selected' by the high affinities of the antibodies encoded against a foreign antigen. The mRNA of these genes then converts immunoglobulin genes in the germline.[29] This may occur via transduction of the mRNA (see Chapter 10) by endogenous retroviruses followed by reverse transcription of the mRNA into cDNA and insertion of the cDNA into the germline in place of a pre-existing gene with a different sequence.

This direct feedback to the germline is so controversial because it goes against the neo-Darwinian orthodoxy that all genetic variations or mutations arise at random, and that those random mutations are subject to subsequent selection. Moreover, there can be no feedback between soma and germline. However, the weight of evidence is overwhelmingly against the idea that mutations are random, in the sense that they are not directly correlated with the environment. Similarly, feedback between soma and germline is simply not an issue with the majority of organisms that do not have segregated germ cells, so somatically acquired DNA modifications are directly inherited. For organisms with some degree of soma-germline segregation, it may be that the feedback channel via reverse transcriptase specifically serves that function.

Adaptive mutations to order

The first indications that mutations are neither random nor rare were obtained when populations of *E. coli* were plated on media containing high concentrations of a metabolite that they cannot use, together with very low concentrations of a carbon source that can keep them alive, but non-growing. Under such conditions, the starving bacteria mutate at high rates, apparently specifically in genes that subsequently enable them to utilise the metabolite, and hence to grow. These mutants do not pre-exist, but arise only *after* the cells are plated, and they do not arise unless the metabolite is present.

One of the first experiments involved the use of mutant strains of *E. coli* in which a gene that breaks down lactose, β-galactosidase, had been deleted, so it could not metabolise the sugar. It was plated on a medium with high concentrations of lactose and some minimal carbon source. When the other carbon source was exhausted, mutant colonies began to appear which possessed lactose-splitting activity. The enzyme responsible was not the β-galactosidase that had been deleted, but another enzyme altogether, called *ebg* (evolved β-galactosidase) mapping to the opposite side of the bacterial genome.[30] This had undergone mutations which gave it lactose metabolising activity. By itself, this result was unremarkable, as it could be explained away in terms of the artificial selection of a fortuitous variation. However, the experiment was repeated by other geneticists,[31] who isolated 34 independent lactose-utilising strains by the same method. All of these contained enzyme activity identical to *ebg*. Moreover, in 31 of the strains, the synthesis of the newly-evolved enzyme was regulated by lactose, i.e., there must have been a mutation in another gene which interacted with lactose to regulate *ebg*. The chance of two random mutations arising in the manner suggested by orthodox neo-Darwinian theory was 10^{-18}. This means one would have to grow about one hundred thousand litres of bacterial culture - the size of a football stadium - to get *one* of these mutants arising by chance.

Many subsequent experiments have confirmed that these mutations arise at frequencies many orders of magnitude above the 'spontaneous', 'random' mutation rate, which is about 10^{-9} or less. These findings led John Cairns and his colleagues to suggest that "bacteria in stationary phase, have some way of producing (or selectively retaining) only the most appropriate mutations"[32] The phenomenon has since been referred to as 'directed mutations' or 'adaptive mutations'.

The phenomenon clearly exists, not only in bacteria but in yeast cells, and possibly also in fruit flies, according to a recent report.[33] But there does not seem to be one single mechanism responsible. DNA replication appears to be necessary but, beyond that, a variety of mechanisms may be involved, including hypermutations due to ineffective DNA repair, genetic recombination, and single base-pair deletions. Adaptive mutations in starving bacteria are due to a spectrum of genetic changes distinct from the 'spontaneous', random ones.[34] Despite this, neo-Darwinists are still trying very hard to salvage the 'randomness' hypothesis by saying that random errors are generated which are subsequently selected. But, as all the cells in the population are genetically uniform, and all the cells are capable of generating those variations, there can be no question of selection in the usual neo-Darwinian sense. Instead it is a physiological response of starving cells involving genetic changes that enable them to overcome the starvation.

In a recent review, molecular geneticist James Shapiro offers an appropriate perspective of genomic fluidity in bacteria that sums up the discoveries of molecular genetics over the past 25 years. He states that bacteria may have "little tolerance for purely random variability". Bacterial cells possess a wide range of repair and proofreading functions to remove accidental change to DNA sequences and correct errors resulting from physiological insults. But the same cells also possess numerous biochemical mechanisms for changing and reorganising DNA, suggesting that such changes are the consequence of 'natural genetic engineering' which enables bacteria to respond to environmental challenges. The ability to activate these mechanisms under stress can significantly accelerate evolutionary change in crisis without threatening genetic stability in ordinary circumstances. He feels it is not too unreasonable to predict that further studies of mobile genetic elements and DNA rearrangements will ultimately uncover truly directed mutations. Just as signal-transduction systems of the cell can direct the transcriptional apparatus to specific locations in the genome to express specific genes, so mutational apparatus(es) can be similarly directed to mutate specific genes.

Notes

1. Much of this account is based on Ho and Goodwin, 1987; Ho, 1987a,b; see also Jones and Taylor, 1995.

2. See Ho, 1987b.

3. See Landman, 1991; Jablonka and Lamb, 1995; also Ho, 1988b.

4. Blanc *et al*, 1996; Maier *et al*, 1996.

5. Lau *et al*, 1997; O'Connell *et al*, 1997.

6. Blanc *et al*, 1996.

7. O'Connell *et al*, 1997,

8. Lau *et al*, 1997.

9. See Ho, 1993, 1997c; also Laszlo, 1994, 1996.

10. The account which follows is based on Ho, 1987b.

11. Dover and Flavell, 1982, backcover.

12. Gierl, 1990.

13. Temin, 1980.

14. Baltimore, 1985.

15. See Flavell, 1982.

16. Temin and Engels, 1984.

17. See Rennie, 1993.

18. Gierl, 1990.

19. McClintock, 1984.

20. Temin and Engels, 1984.

21. Bostock and Tyler-Smith, 1982; Gudkov and Kopnin, 1985.

22. See Cullis, 1983, 1988; also Ho, 1987a,b.

23. See Dover, 1982; also Ho, 1987b.

24. See Jablonka and Lamb, 1995; also Ho, 1996d.

25. Ho *et al*, 1983.

26. Temin, 1980.

27. Soares *et al*, 1985.

28. Hattori *et al*, 1986.

29. Rothenfluh and Steele, 1993; Rothenfluh *et al*, 1995.

30. Campbell *et al*, 1973; see also Ho, 1987a.

31. Hall and Hartl, 1974.

32. Cairns *et al*, 1988.

33. Riede, 1996.

34. See Foster, 1992; Longerich *et al*, 1995.

35. See Shapiro, 1997.

Chapter 9

Perils amid Promises of Genetically Modified Foods

Agricultural biotechnology cannot alleviate the existing food crisis. On the contrary, it is inherently unsustainable and hazardous to biodiversity, human and animal health. A drastic change of direction will be required to support conservation and sustainable development of indigenous agricultural biodiversity, which both satisfies the stated aims of the Biodiversity Convention and guarantees long-term food security for all.

The Food 'Crisis'

By the year 2000, the world will need to consume over 2 billion tonnes a year of wheat, rice, maize, barley and other crops - an increase of 25% compared with 1995 figures.[1] This view was echoed by the World Bank Report for the 1996 World Food Summit in Rome,[2] which warned that the world would have to double food production over the next 30 years. One major solution on offer to 'feed the world' is agricultural biotechnology. This, it has been proposed, could be used to genetically modify crops for herbicide, pest, and disease resistance, to improve nutritional value and shelf-life, and also, for the future, to bring about promises of drought and frost resistance, nitrogen fixation[3] and increased yield.[4]

Agricultural biotechnology is big business, and the mission to feed the world has the irresistible ring of a noble obligation. The same goes for improving the nutritional value of foods. Despite prices having dropped to the lowest on record, more than 800 million people still go hungry,[5] and 82 countries - half of them in Africa - neither grow enough food, nor can afford to import it. Infant mortality rates - a sensitive indicator of nutritional stress - have been experiencing an upturn in recent years, reversing a long-term historical trend. Large numbers of children suffer from malnutrition in developing countries. In India alone, 85% of children under five are below the normal, acceptable state of nutrition.[6]

In view of the current crisis in food production, and the support for agricultural biotechnology as a solution to the crisis - as expressed by the World Bank Report and by Chapter 16 of Agenda 21 of the United Nations Convention on Biological Diversity (see Chapter 2) - it is all the more important to examine the major claims and promises of the technology, as well as the uncertainties and hazards which are not adequately taken into account in existing practices and regulations.

Can Genetically Modified Foods feed the World or Improve Nutrition?

The poverty trap of unequal power relationships

Under-nutrition and malnutrition, found everywhere in the developing as well as the developed world, stem from poverty, as was admitted in the World Bank Report.

In the Third World, poverty was created, in large measure, by centuries of colonial and post-colonial economic exploitation under the free-trade imperative, and has been exacerbated since the 1970s by the introduction of the intensive, high-input industrial agriculture of the Green revolution.[7] The concentration on growing crops for export has benefited the corporate plant breeders and the elite of the Third World at the expense of ordinary people. In 1973, thirty-six of the nations most seriously affected by hunger and malnutrition exported food to the U.S. - a pattern that continues to the present day.[8] The 'liberalisation of trade' under the current World Trade Organisation agreement will make things much worse.[9] While Southern countries are obliged to remove subsidies to their farmers, subsidies to Northern producers have remained untouched. This unequal competition will deprive millions of peasants of their livelihood. In addition, as part of the same WTO agreement, the intellectual property rights of corporate gene manipulators in the North will be protected (see Chapter 2), and that will restrict the use of indigenous varieties that were previously freely cultivated and sold. Thus, seeds protected by patents will no longer be able to be saved by farmers for replanting without annual royalties being paid to the company which owns them.

Another factor already adversely affecting agricultural biodiversity in Europe is the Seed Trade Act which makes it illegal to grow and sell non-certified seeds, produced by organic farmers from indigenous varieties, certification being biased towards the commercial varieties currently used in agricultural biotechnology.[10] Far from providing cheaper food for all, agricultural biotechnology will further undermine the livelihoods of small organic farmers all over the world, resulting in increased loss of indigenous agricultural biodiversity.

Discussions on food supply are invariably linked to population growth in the Third World. But these discussions leave out the unequal power relations which exist between different countries and different groups of people. 'Food scarcity', like 'overpopulation' are both socially generated. While populations in the North are suffering from obesity, cardiovascular diseases and diabetes from over-consumption, populations in the South are dying of starvation. Simplistic 'solutions' which leave out the unequal power relations are oppressive, and ultimately "reinforce the very structures creating ecological damage and hunger"[11]

Biological diversity, food security and nutrition

Biological diversity and food security are intimately linked. Communities everywhere have derived livelihoods from natural diversity in wild and domesticated forms. Diversity is the basis of ecological stability.[12] Recent studies show that diverse ecological communities are more resilient to drought and other environmental disturbances which cause the population of individual species to fluctuate widely from year to year.[13] Species within an ecological community are interconnected in an intricate web of mutualistic as well as competitive interactions, of checks and balances that contribute to the survival of the whole (see Chapter 14). This has important implications for *in situ* conservation, particularly at a time when it is being estimated that 50,000 species will go extinct every year over the next decade.[14]

The same principles of diversity and stability operate in traditional agriculture.[15] Throughout the tropics, traditional agroforestry systems commonly contain well over 100 annual and perennial plant species per field. A profusion of varieties and land races are cultivated which are adapted to different local environmental conditions and possess a range of natural resistances to diseases and pests. Spatial diversity through mixed cropping is augmented by temporal diversity in crop rotation, ensuring the recycling of nutrients that maintain soil fertility. These practices have effectively prevented major outbreaks of diseases and pests and buffered food production from environmental exigencies.

The diversity of agricultural produce is also the basis of a balanced nutrition. Nutrition not only depends on the right balance of protein, carbohydrates and fats, but also on a combination of vitamins, essential metabolites, cofactors, inorganic ions and trace metals, which only a varied diet can provide. A major cause of malnutrition world-wide is the substitution of the traditionally varied diet for one based on monoculture crops. The transfer of an exotic gene into a monoculture crop can do little to make up for the dietary deficiencies of those suffering from monoculture malnutrition. The nutritional value of beans, or a combination of rice and beans, will always be greater than that of the transgenic rice with a bean gene.

Monoculture and transgenic threats

It is now indisputable that monoculture crops introduced since the Green revolution have adversely affected biodiversity and food security all over the world. According to a FAO report, by the year 2000 the world will have lost 95% of the genetic diversity utilised in agriculture at the beginning of this century.[16] Monoculture crops are genetically uniform and, therefore, notoriously prone to disease and pest outbreaks. The corn belt of the United States was last devastated by corn blight in 1970-1971, while, in 1975, Indonesian farmers lost half a million acres of rice to leaf hoppers. Genetic modification for disease or pest resistance will not solve the problem, as intensive agriculture itself creates the conditions for new pathogens to arise.[17] In 1977, a variety of rice, IR-36, created to be resistant to 8 major diseases and pests, including bacterial blight and tungro, was nevertheless attacked by two new viruses called 'ragged stunt' and 'wilted stunt'. Thus, not only do new varieties have to be substituted every three years, they require heavy input of pesticides to keep pests at bay.

The high inputs of fertilisers, water, pesticides and heavy mechanisation required by monoculture crops have had devastating environmental effects.[18] Teddy Goldsmith, who started the ecology movement in Britain in 1970, has been a long-time critic of global financial institutions such as the World Bank and the International Monetary Fund for the anti-ecological projects they finance in the

Third World, such as the construction of big dams for irrigation and roads which hasten the clearing of forests Between 1981 and 1991, the world's agricultural base fell by some 7 percent, primarily due to environmental degradation and water shortages. One-third of the world's croplands suffers from soil erosion, which could reduce agricultural production by a quarter between 1975 and the year 2000. In India, 800,000 square kilometres are affected, with many areas turning into scrub or desert. Deforestation has resulted in 8.6 million hectares of degraded land in Indonesia, which is unable to sustain even subsistence agriculture. Throughout the tropics, vast areas are vulnerable to flooding. Of the world's irrigated land, one-fifth - 40 million hectares - suffers from waterlogging or salination. The resultant pressures on agricultural land led to the further marginalisation of small farmers, swelling the ranks of the dispossessed and hungry, while indigenous natural and agricultural biodiversity are eliminated at accelerated rates.

Transgenic crops are created from the same high-input monoculture varieties of the Green revolution, and are likely to make things worse. The greatest proportion of transgenic crop plants is now engineered to be resistant to herbicide, with companies engineering resistance to their own herbicide to increase sales of herbicides with seeds.[19] The immediate hazard from herbicide resistant crops is the spread of transgenes to wild relatives by cross-hybridisation, creating super-weeds. Herbicide-resistant transgenic oilseed rape, released in Europe, has now hybridised with several wild relatives.[20]

There are yet other problems. Herbicide resistant transgenic crops make it possible to apply powerful herbicides, killing many species, directly onto crops. This is so for Monsanto's Roundup, which is lethal to most herbaceous plants. The U.S. Fish and Wildlife Service has identified 74 endangered plant species threatened by the use of herbicides like glyphosate.[21] This product reduces the nitrogen-fixing activity of soils and is toxic to many species of mycorrhizal fungi which are vital for nutrient recycling in the soil. Glyphosate-type compounds are the third most commonly-reported cause of pesticide illness among agricultural workers. The use of this highly toxic non-discriminating herbicide will lead to the large-scale elimination of indigenous species and cultivated varieties, damaging soil fertility and human health besides. Herbicide-resistant transgenic crops also become weeds in the form of 'volunteer plants' germinated from seeds after the harvest, so that other herbicides then have to be applied in order to eliminate them, with yet further impact on indigenous biodiversity.

Food security depends on agricultural biodiversity

In order to counteract the crisis of environmental destruction, loss of agricultural land and indigenous biodiversity created by decades of intensive farming, there has been a global move towards holistic, organic farming methods that revive traditional practices. Previous promoters of the Green revolution are now calling for a shift to sustainable agriculture. Sustainable agriculture is promoted in Chapter 14 of Agenda 21 of the United Nations Convention on Biodiversity, signed by more than 140 countries. Large-scale implementation of bio-dynamic farming and sustainable agriculture is succeeding in the Philippines.[22] In Latin America, a number of non-government organisations have joined forces to form the Latin American Consortium on Agroecology and Development, to promote agroecological techniques which are sensitive to the complexities of local farming methods. Programmes introducing soil conservation practices and organic fertilisation methods tripled or quadrupled yields within a year.[23] Successive studies have highlighted the productivity and sustainability of traditional peasant farming in the Third World[24] as well as in the North, according to a report published by the U.S. National Academy of Sciences.[25]

Many, if not all, southern countries still possess the indigenous genetic resources - requiring no further genetic modification - that can guarantee a sustainable food supply.

"Over centuries of agricultural practice, traditional societies have developed an incredible variety of crops and livestock. Some 200-250 flowering plants species have been domesticated, and genetic diversity amongst each of these is astonishing: in India alone, for instance, farmers have grown over 50,000 varieties of rice *Oryza sativa*. In a single village in north east India, 70 varieties are being grown... farmers (especially women) repeatedly used and enhanced some varieties which were resistant to disease and drought and flood, some which tasted nice, some which were coloured and useful for ritual purposes and some which were highly productive."[26]

In Brazil, hundreds of rural communities in the north-east are responding to the current crisis in food production by organising communal seed banks to recover traditional indigenous varieties and to promote sustainable agricultural development, with little or no government support.[27]

It is significant that the World Bank is reported to be planning sharp changes in policy to concentrate its efforts on small farmers in developing countries.[28] It seems obvious that, in order to guarantee long-term food security and feed the world, we can do no better than take the aim of the Convention on Biological Diversity to heart, i.e. help to conserve and sustain existing indigenous agricultural diversity world-wide, and to develop this diversity as the basis of a secure and nutritious food base for all.

Thus, there is no need for genetically modified crops. On the contrary, they will undermine food security and biodiversity. Under the combined effects of monopoly of transnational genetic manipulators' intellectual property rights and 'free trade' agreements of the World Trade Organisation, the livelihoods of small farmers will be further compromised, both by seed royalties and the restrictive practices of seed certification, and unfair competition from subsidised Northern produce. At the same time, the use of toxic, wide-spectrum herbicides with herbicide-resistant transgenic crops will result in irretrievable losses of indigenous agricultural and natural biological diversity.

There are, in addition, problems and hazards inherent to the practice of the technology itself, which make the regulation of the technology, by a legally binding international Biosafety Protocol under the Convention of Biodiversity, a matter of urgency.

Agricultural Biotechnology is misguided by Wilful Ignorance of Genetics

In a publication which aims to "provide consumers with clear and comprehensible information about products of the new [bio]technology", we are told that: "Research scientists can now precisely identify the individual gene that governs a desired trait, extract it, copy it and insert the copy into another organism. That organism (and its offspring) will then have the desired trait.."[29]

This reaffirms the genetic determinist idea that one gene controls one character trait, and that transferring the gene results in the transfer of the corresponding trait to the genetically modified organism, which can then pass it on indefinitely to future generations. It presents the process of genetic modification as a precise and simple operation.

The above account - so typical of that found in publications promoting 'public understanding' - is based on a simplistic assumption of genetics that both classical geneticists and plant breeders have rejected for many years, and which has been thoroughly invalidated by all the research findings in the new genetics (see Chapters 3 and 8). Unfortunately, most molecular geneticists, apart from being absorbed into industry, also lack training in classical genetics, and suffer from a severe molecular myopia that prevents them from appreciating the implications and broader perspective of the findings in their own discipline. Damages from intensive agricultural practices have indeed come about because they are based on the old reductionist paradigm, as Vandana Shiva has argued so convincingly.[30] For the same reason, agricultural biotechnology will bring new problems and hazards, I have already pointed out some of its failings in Chapter 3. It will be instructive to go into further details here.

Dangers of Ignoring the Interconnected Genetic Network

Because no gene ever functions in isolation, there will almost always be unexpected and unintended side-effects from the gene or genes transferred into an organism.

One major concern over transgenic foods is their potential to be toxic or allergenic, which has become a concrete issue since a transgenic soybean containing a brazil-nut gene was found to be allergenic.[31] Recent studies suggest that allergenicity in plants is connected to proteins involved in defence against pests and diseases. Thus, transgenic plants engineered for resistance to diseases and pests may have a higher allergenic potential than unmodified plants.[32]

New proteins from bacteria, such as the Bt toxin currently engineered into many transgenic crops, cannot be tested for allergenicity because allergic reactions depend on prior exposures. *This means that post-market monitoring and clear segregation and labelling of transgenic products are essential for proper consumer protection.* Most identified allergens are water-soluble and acid-resistant. Some, such as those derived from soya, peanut and milk, are very heat-stable, and are not degraded during cooking, whereas fruit-derived allergenic proteins are heat-labile.[33]

A transgenic yeast was engineered for increased rate of fermentation with multiple copies of one of its own genes, which resulted in the accumulation of the metabolite, methylglyoxal, at toxic, mutagenic levels.[34] This case should serve as a warning against applying the 'familiarity principle' or 'substantial equivalence' in risk assessment. We simply do not have sufficient understanding of the principles of physiological regulation to enable us to categorise, *a priori*, those genetic modifications that pose a risk and those that do not.

I wrote to the UK Ministry of Agriculture, Fisheries and Foods (UK MAFF) of my concerns over the inadequacy of regulation of biotechnology. In answer to my query on the possibilities of toxic or allergenic effects due to new transgene products and products resulting from interactions with host genes, their experts replied,

"As part of ACNFP (Advisory Committee on Novel Food Products) evaluations of GMOs and their products, companies are required to produce data from tests carried out by independent laboratories to address these concerns. Data must be supplied on the safety of the transgene product itself as well as the safety of the food containing the transgene and, if present, its product. The ACNFP requests detailed descriptions of transgenes and their origins and information about the entire DNA sequence that has been transferred into the host organism. When 'substantial equivalence' has been claimed - for an oil from oilseed rape, for example - the ACNFP asks for screening to be carried out on a number of samples for any toxic components known to be produced by the host to ensure that levels have not increased as a result of the genetic modification..."

In other words, there are no data required on any previously *unknown* products that may have been produced by gene interactions. Furthermore, the claim of 'substantial equivalence' is sufficiently vague to make regulation even weaker than it appears.[35]

If one takes seriously the highly interconnected network in which genes function, then a proper risk assessment ought to include a complete profile of the proteins produced, as well as the pattern of metabolites. Both of those involve laboratory methods that are now routine. Labelling of products is indispensable if the long-term effects on consumers ingesting new bacterial proteins are to be monitored, as they must be, for reasons I have stated above. The European Commission on Agriculture has, belatedly, called for a complete segregation and labelling of transgenic products.

Danger of Ignoring the Ecology of Genes and Organisms

Single genes impacts on the ecosystem

The most immediate and easily observable impacts of transgenic plants on the ecological environment are due to cross-pollination between transgenic crop-plants and their wild relatives to generate super-weeds. Field trials have shown that cross-hybridisation has occurred between herbicide-resistant transgenic *Brassica napa* and its wild relatives; *B. campestris,*[36] *Hirschfeldia incana,*[37] and *Raphanus raphanistrum.*[38] These impacts have been predicted by ecologists such as Rissler and Mellon,[39] and arise from the introduction of any exotic species, whether genetically engineered or not.

Impacts which are generally underestimated are those due to transgenic soil bacteria. As very few molecular geneticists have any training in soil ecology, they will be ignorant of the important role played by the soil microbes in recycling nutrients for the growth of crop plants. Soil microbiologists Elaine Ingham and her student tested a common soil bacterium, *Klebsiella planticola*, engineered to produce ethanol from crop waste, in jars containing different kinds of soil in which a wheat seedling had been planted.[40] The experiments showed that, in all soil types, the growth of the wheat seedling was drastically inhibited. This was due to the ethanol produced, which had adverse effects on different microbes that were involved in recycling nutrients for the wheat seedling. Elaine has talked about this in several TWN-sponsored seminars, to great effect. She and her colleagues now run a consultancy and research firm for organic farming in the U.S., which is a marvellous way to resist the agrochemical biotechnological encroachment.

The instability of transgenic lines

Traditional breeding methods involve crossing closely-related varieties or species containing different forms of the same genes. Selection is then practised over many generations under field conditions, so that the desired characteristics and the genes influencing those characteristics, *in the appropriate environment*, are tested and harmonised for stable expression over a range of genetic backgrounds. Different genetic combinations, moreover, will perform differently in different environments. This genotype-environment interaction is well-known in traditional breeding, so it is not possible to predict how a new variety will perform in untested environments. In many cases, new varieties will lose their characters in later generations as genes become shuffled and recombined, or as they respond to environmental changes.

In the new genetic modification, completely exotic genes are often introduced into organisms. In the case of plants, the genes are often introduced into plant cells in tissue culture, and transgenic plants are regenerated from the cells after selection in culture. The procedures inherently generate increased genetic instability in the resulting transgenic line.

First, the tissue culture technique itself introduces new genetic variations at high frequencies. These are known as *somaclonal variations.*[41] That is because the cells are removed from the internal, physiological environment of the plant which stabilise their gene expression and genetic complement *in vivo* (see Chapter 14). It is part of the spectrum of ecological interactions between organism and environment that keep gene expression, genes and genome structure stable in the organism as a whole. Unilever used tissue culture techniques to regenerate oil palms for planting in Malaysia several years ago. This practice has now been abandoned as many plants aborted in the field or failed to flower.[42] The second reason for increased instability of transgenic lines is that the process of gene insertion is random and a lot of secondary genetic effects can result, as mentioned earlier. Third, the extra DNA integrated into the transgenic organism's genome disrupts the structure of its chromosome, and can itself cause chromosomal rearrangement,[43] further affecting gene function. Finally, all species have cellular mechanisms which tend to eliminate or inactivate foreign DNA.[44] Transgene instability, particularly 'gene-silencing' [45] - the inability of the introduced gene to become expressed in subsequent generations - has been discovered only within the past few years, and is now a recognized problem in both farm animals and plants.[46] In transgenic tobacco, 64% to 92% of the first generation of transgenic plants become unstable. Similarly, the frequency of transgene loss in *Arabidopsis* ranges between 50% and 90%. Instability arises both during the production of germ cells and in cell division during plant growth. The commonest cause is gene silencing due to the chemical modification of the introduced DNA by methylation - a reaction adding a methyl group, $-CH_3$, to the base cytosine or adenosine. Other causes are due to DNA rearrangements and excision of the transgene. The long-term agronomic viability of transgenic crops has yet to be proven. Calgene's Flavr Savr tomato, engineered for improved shelf-life, was a financial disaster (as was the transgenic strawberry).[47] Apart from side-effects, such as a skin too soft for the tomato to be successfully shipped, it also failed to grow in Florida, as it was created in California. At least in that regard, commercialisation had been premature. In 1996/7, Monsanto's transgenic Bt-cotton crop, engineered to be resistant to the cotton bollworm, failed to live up to its promise in the field in both the USA and Australia, partly on account of transgene inactivation.[48] Farmers should beware.

By contrast, the long-established indigenous local varieties and land races are the most stable, as genes and environment have mutually adapted to reinforce the stable expression of desirable characteristics for hundreds, if not thousands, of years. There is no quick fix to establishing ecological balance, which must be restored in order to guarantee our long-term food security.

Empty promise of 'high-yielding' and 'nitrogen-fixing' crops

In the light of all this, it is irresponsible to claim that genetic modification can make high-yielding or nitrogen-fixing transgenic plants. Yield is a complex character - a polygenic trait (see Chapter 6) - dependent on many still largely unknown genes as well as on environmental conditions.[49] Furthermore, it cannot be identified, and hence cannot be selected for in tissue culture. Hopes of discovering individual 'genetic markers' for yield are unrealistically optimistic. A complex character such as yield cannot be 'transferred' by transferring one or two genes. Even if all the genes required could be transferred, the problems of genetic instability would only be correspondingly multiplied.

The same goes for nitrogen fixation, which refers to the ability of a relatively small number of bacterial species to reduce atmospheric nitrogen to ammonia, a product that can be used by plants and other microbes to make amino acids and, hence, proteins and other nitrogenous compounds essential for life.[50] On a global scale, the amounts of nitrogen fixed by these bacteria are in the region of 200 million tonnes each year. This is compared to the current Haber-Bosch chemical nitrogen-fixing industrial process, which accounts for 40 million tonnes of nitrogen fertiliser synthesised each year. Of especial importance are nitrogen-fixing bacteria of the genus *Rhizobium* which live in symbiotic relationship in the roots nodules of legumes.

Shifting from chemical to biologically fixed nitrogen is desirable, as much of the chemical fertiliser applied to soil is leached from it, polluting drinking water, and leading to algal blooms in the water system, annoxia, and hence a decline in fish and shellfish populations.

However, the most important nitrogen-fixing bacteria work in symbiotic relationship with higher plants. The reason is that the process is thermodynamically uphill and extremely costly in terms of energy. It depends on at least 17 genes in the bacterium, and 50 genes in the plant.[51] Moreover, it has to take place in the absence of oxygen, which is toxic to the major nitrogen-fixing enzyme complex. Most serious molecular geneticists do not rate as realistic the prospect of biotechnology creating transgenic nitrogen-fixing crop-plants. However, they recognize the value of traditional breeding methods in improving the efficiency of the already existing symbiotic relationship between nitrogen-fixing bacteria and higher plants and, hence, reducing the need for chemical fertilisers.

Dangers of Ignoring Fluid Genome Process

Bt-resistance

Pesticide resistance has been a major and persistent problem in intensive agriculture. The rapid evolution of insecticide resistance has become a textbook example of the supposed power of neo-Darwinian natural selection to increase the frequency of 'rare random mutations' that confer resistance. Actually, insecticide resistance has turned out to be due to genetic changes that can occur in a large proportion, if not all, individuals in insect populations, including mosquitoes, houseflies and aphids exposed to sub-lethal levels of insecticide, and this has been known for more than ten years. Resistance often involves the amplification of genes-encoding enzymes that detoxify the chemical, and is part and parcel of the 'fluid genome' mechanisms common to all cells challenged with toxic substances - including anti-cancer drugs in mammalian cells and antibiotics in bacteria (see Chapters 8 and 11). Resistance to glyphosate readily arises in plant cells lines exposed to the herbicide, and involves amplifications of detoxifying genes.[52] In the light of this knowledge, one could have predicted that transgenic plants with built-in insecticide would favour the acquisition of resistance by insect populations, which are effectively exposed continuously at sublethal levels.

A range of the insecticidal δ-endotoxins, or *Bt*-toxins, made by genes isolated from the soil bacterium *Bacillus thuringiensis*, have been introduced into many transgenic crops, including cotton, maize and potato, and field-tested at least since 1990.[53] These genes encode a large protein *pro*-toxin. The pro-toxin is not harmful to insects by itself, but must be processed in the gut of susceptible species of insect larvae - which possess the processing enzyme - into the toxin that kills the larvae. By modifying the gene and incorporating aggressive promoters and enhancers to boost gene expression, high levels of the protein can be expressed in transgenic plants. In some transgenic plants, a truncated form of the gene is used so that no processing by the susceptible insect is required to generate the toxin. It is, therefore, completely non-selective, and will harm non-target insects that do not have the enzymes to process the pro-toxin, as well as the pests for which it is intended.

Bt sprays had previously been used for 40 years by organic farmers as an environmentally friendly biopesticide to control pests. But the new generation of transgenic crops has already generated an ecological crisis in creating Bt resistances among major insect pests in the U.S. where these transgenic crops have been released over the past two years.[54] Bt resistance contributed substantially to the problems experienced by the Bt-cotton crop in USA and Australia in 1996/7. Strategies for crisis management had to be adopted, such as supplementary sprays, the creation of non-transgenic 'refugia' to continue breeding non-resistant insect pests, the promise of transgenic plants engineered with multiple toxins all at once, and an intensified search for new toxins.[55] Unfortunately, the resistance trait is highly stable, and also exhibits broad spectrum cross-resistance to other delta-endotoxins, which undermines many potential options for resistance management.[56]

These measures betray the short-termism of reductionist, non-ecological thinking. From past experiences, it can be predicted that newer and more powerful resistances and multiple resistances will be acquired by insect pests. This continual warfare with nature has already been shown to fail, as losses due to insects are currently estimated as accounting for 20% to 30% of total production, at a time when they are being fought with a deadly arsenal of chemical pesticides. The new generations of transgenic plants with biopesticide genes are even more dangerous. They may be destroying the last stronghold of the ecosystem's ability to readjust and rebalance itself in the face of the assaults of intensive agriculture. The biopesticides have all been isolated from soil bacteria. It does not take a great deal of imagination to infer that they probably play an indispensable role in natural pest control. If, as the result of commercial releases of biopesticide-producing transgenic plants, insect pests develop resistances on a large scale, there will be nothing left for the ecosystem to fall back on. The future of agriculture within such a scenario may be among the first of the genetically engineered nightmares.

Hazards from Horizontal Gene Transfer and Recombination

The most underestimated hazards of agricultural biotechnology are from horizontal gene transfers (see Chapters 8 and 10). There is now abundant evidence that gene transfer vectors mediate horizontal gene transfer and recombination, spreading antibiotic resistance and generating new pathogens. Antibiotic resistance arose as the result of the profligate use of antibiotics in intensive farming, which predates genetic engineering. However, current transgenic plants often contain antibiotic-resistance marker genes. When released into the environment, these genes will exacerbate the spread of antibiotic resistance. This consideration is behind the UK's initial rejection of Ciba-Geigy's transgenic maize which contains the marker gene for ampicillin resistance,[57] as ampicillin is still much in use. Nevertheless, the UK has authorised the marketing of Zeneca's transgenic tomato paste as well as Calgene's transgenic tomato, both of which carry the marker gene for kanamycin resistance. Although it is claimed that kanamycin is no longer in use, as it has already been supplanted by new generations of aminoglycoside antibiotics, at least one kanamycin resistance gene used as a genetic marker has been found to confer cross-resistance to two new generation aminoglycosides, amikacin and tobramycin.[58]

Spreading genes via the natural microbial populations

There is an obvious route for the vectors containing transgenes in transgenic higher plants and animals (transgenic fish and shellfish) as well as microorganisms to take in order to spread - and that is via the teeming microbial populations. Microbial populations in all environments form large reservoirs supporting the multiplication of the vectors, enabling them to spread to all other species. They also provide an opportunity for the genetic elements to recombine with other viruses and bacteria and generate new genetic elements and pathogenic strains of bacteria and viruses which will, at the same time, be antibiotic resistant.

This route cannot be ignored even for higher organisms, as horizontal transfers of transgenes and marker genes *have* been experimentally demonstrated in the laboratory: from transgenic potato to a bacterial pathogen, and between transgenic plants and soil fungi under co-cultivation. In the experiments with transgenic potato, a high 'optimal' gene transfer frequency of 6.2×10^{-2} was observed by a researcher in the laboratory, from which, using dubious assumptions, an extremely low frequency of 2.0×10^{-17} was 'calculated' under 'natural idealised conditions'.[59] It is impossible to know the precise frequencies for such horizontal gene transfer under natural conditions as very few actual studies have been carried out. Nor has horizontal gene transfer been seriously monitored in previous field trials or releases. Many transgenic plants have been made with a vector constructed from the *Agrobacterium* Ti (tumour-inducing) plasmid, which is already known to be able to integrate itself into plant cells (see Chapter 10). It is therefore possible for the integrated vector to undergo secondary mobilisation and to end up in other plants, either via microbes in the environment, or via insect vectors. In view of the already documented propensity for horizontal gene transfer, great caution must be exercised to avoid releasing undesirable transgenes and marker genes into the environment.

Viral resistance transgenes generate live viruses

A major class of transgenic plants are now engineered for resistance to viral diseases by incorporating the gene for the virus's coat protein. Some molecular geneticists have expressed concerns that transgenic crops engineered to be resistant to viral diseases with genes for viral coat proteins might generate new diseases by several known processes. The first, *transcapsidation*, has already been detected, and involves the DNA/RNA of one virus being wrapped up in the coat protein of another so that viral genes can get into cells which otherwise exclude them. The second possibility is that the transgenic coat protein can help defective viruses multiply by *complementation*. The third possibility, *recombination*, has been demonstrated in an experiment in which *Nicotiana benthamiana* plants, expressing a segment of a cowpea chlorotic mottle virus (CCMV) gene, were inoculated with a mutant CCMV, missing that gene. The infectious virus was indeed regenerated by recombination.[60] There is now also evidence that transgenic plants increase the frequency of viral recombination, owing to the continual expression of the viral coat protein gene.[61] As plant cells are frequently infected with several viruses, recombination events will occur and new and virulent strains will be generated. Viral recombination is well documented in animals and the resulting recombinant viruses are strongly implicated in causing diseases (see Chapter 13). As in animals, plant genomes also contain many endogenous proviruses and related elements which can potentially recombine with the introduced transgene.

Another strategy for viral resistance made use of benign viral 'satellite RNAs' as transgenes, thereby attenuating the symptoms of viral infection. However, these were found to mutate to pathogenic forms at high frequencies.[62] These already documented pathogenic recombinants and mutants, regenerated from viral resistant transgenic plants, are particularly significant, as viruses are readily transmitted from one plant to another by many species of aphids and other insects that attack the plants. There is a distinct possibility of new broad-range recombinant viruses arising, which could cause major epidemics.

A potentially major source of new viruses arising from recombination has been pointed out by molecular geneticist, Joe Cummins.[63] This is the powerful promoter gene from cauliflower mosaic virus (CaMV), which is routinely used to drive gene expression in transgenic cropplants for herbicide or disease resistance. Like the viral coat-protein gene, this viral gene can also recombine with other viruses to generate new broad range viruses. The CaMV has sequence homologies to human retroviruses such as the AIDS virus, human leukaemic virus and human hepatitis B virus, and the promoter gene can drive the synthesis of these viruses as well. There is thus a possibility for the CaMV promoter to recombine with human viruses when ingested in food (see below).

Vectors can infect mammalian cells and resist breakdown in the gut

Among the important factors to consider in the safety of transgenic organisms used as food are the extent to which DNA, particularly vector DNA, can resist breakdown in the gut, and the extent to which it can infect the cells of higher organisms.

Studies made since the 1970s have documented the ability of bacterial plasmids carrying a mammalian virus to infect cultured mammalian cells, which then proceed to synthesise the virus, even though no eukaryotic signals for reading the genes are contained in the plasmid. This is because endogenous provirus and other elements can provide helper-functions which are missing. Similarly, bacterial viruses or baculovirus can also be taken up by mammalian cells.[64] Baculovirus is so effectively taken up by mammalian cells that it is now being developed as a gene transfer vector in human gene replacement therapy. At the same time, baculovirus is genetically engineered to kill insects more effectively, with genes encoding diuretic hormone, juvenile hormone, Bt endotoxin, mite toxins and scorpion toxin. The recombinant virus is sprayed directly onto crop-plants.[65] Recently, a recombinant baculovirus has even been made containing an anti-sense gene from a human cancer gene, *c-myc*. So, what happens when humans eat foods containing vectors and viral sequences?

It has long been assumed that our gut is full of enzymes which can rapidly digest DNA. In a study designed to test the survival of viral DNA in the gut, mice were fed DNA from a bacterial virus, and large fragments were found to survive passage through the gut and to enter the bloodstream.[66] This research group has now shown that ingested DNA end up, not only in the gut cells of the mice, but also in spleen and liver cells as well as white blood cells. "In some cases, as much as one cell in a thousand had viral DNA".[67]

A group of French geneticists found that certain pathogenic bacteria have acquired the ability to enter mammalian cells directly by inducing their own internalisation. They found invasive strains of *Shigella flexneri* and *E. coli* that had undergone lysis upon entering the mammalian cells because of an impairment in cell wall synthesis. The researchers developed these strains as DNA transfer systems into mammalian cells. This transfer was described as "efficient, of broad host cell range and the replicative or integrative vectors so delivered are stably inherited and expressed by the cell progeny."[68] The researchers are totally unable to recognise the tremendous risks to health involved in developing such a vector. These cross-kingdom transfer vectors are extremely hazardous, as are transgenic vaccines constructed in plants and plants viruses, which are chimeras of animal viral genes inserted into plant viruses. These will have an increased propensity to invade cells, recombine with endogenous viruses and proviruses or insert themselves into the cell's genome (see Chapter 13).

Within the gut, vectors carrying antibiotic resistance markers may also be taken up by the gut bacteria, which would then serve as a mobile reservoir of antibiotic resistance genes for pathogenic bacteria. Horizontal gene transfer between gut bacteria has already been demonstrated in mice and chickens and in human beings (see Chapter 10).

In view of all this evidence, it would seem unwise to ingest transgenic foods, as foreign DNA can resist digestion. It can be taken up by gut bacteria, as well as by gut cells, and, through the gut, into the blood stream and other cells. DNA uptake into cells can lead to the regeneration of viruses. If the DNA integrates into the cell's genome, a range of harmful effects can result including cancer. Moreover, one cannot assume, without adequate data, that DNA is automatically degraded in *processed* transgenic foods, such as the Zeneca's tomato paste currently on sale in UK supermarkets, as well as the many foods containing processed trangenic soybean or maize. The public is already being experimented on, *without informed consent.* This is surely against the European BioEthics Convention. Yet, almost nothing can be learned, since it is, at present, impossible to collect relevant data when neither labelling nor post-market monitoring is required.

Checklist of Hazards from Agricultural Biotechnology

As a summary of this somewhat complex chapter, I shall reiterate the arguments on why agricultural biotechnology is unsustainable and poses unique hazards to health and biodiversity.

a. Socio-economic impacts

1. Increased drain of genetic resources from South to North.
2. Increased marginalisation of small farmers due to intellectual property rights, and other restrictive practices associated with seed certification.

3. Substitution of traditional technologies and produce.
4. Inherent genetic instability of transgenic lines resulting in crop failures.

b. Hazards to human and animal health

1. Toxic or allergenic effects due to transgene products or products from interactions with host genes.
2. Increased use of toxic pesticides with pesticide-resistant transgenic crops, leading to pesticide-related illnesses in farm workers, and the contamination of food and drinking water.
3. Vector-mediated spread of antibiotic resistance marker genes to gut bacteria and to pathogens.
4. Vector-mediated spread of virulence among pathogens across species by horizontal gene transfer and recombination.
5. Potential for vector-mediated horizontal gene transfer and recombination to create new pathogenic bacteria and viruses.
6. Potential of vector-mediated infection of cells after ingestion of transgenic foods, to regenerate disease viruses, or for the vector to insert itself into the cell's genome causing harmful or lethal effects including cancer.

c. Hazards to agricultural and natural biodiversity

1. Spread of transgenes to related weed species, creating superweeds (e.g. herbicide resistance).
2. Increased use of toxic, non-discriminating herbicides with herbicide-resistant transgenic plants leading to large-scale elimination of indigenous agricultural and natural species.
3. Increased use of other herbicides to control herbicide-resistant 'volunteers', thus further impacting on indigenous biodiversity.
4. Increased use of toxic herbicides destroying soil fertility and yield.
5. Bio-insecticidal transgenic plants accelerating the evolution of bio-pesticide resistance in major insect pests, resulting in the loss of a bio-pesticide used by organic farmers for years.
6. Increased exploitation of natural bio-pesticides in transgenic plants, leading to a corresponding range of resistant insects, depriving the ecosystem of its natural pest controls and the ability to rebalance itself to recover from perturbation.
7. Vector-mediated horizontal gene transfer to unrelated species via bacteria and viruses, with the potential of creating many other

weed species.

8. Vector recombination to generate new virulent strains of viruses, especially in transgenic plants engineered for viral resistance with viral genes.
9. The vectors carrying the transgene, unlike chemical pollution, can be perpetuated and amplified given the right environmental conditions. It has the potential to unleash cross-species epidemics of infectious plant and animal diseases that will be impossible to control or recall.

Conclusion

The World Bank Report for the 1996 Food Summit advocated sustained support for research to develop new plants and technologies, but it also called for "whole new ways" of addressing the problem of the current food crisis, one of which was to concentrate on helping small farmers.

In this chapter, I have presented the reasons why agricultural biotechnology *cannot* alleviate the existing food crisis. On the contrary, *it is inherently unsustainable, and extremely hazardous to biodiversity, human and animal health.* A drastic change of direction is indeed required, targeted to supporting conservation and sustainable development of indigenous agricultural biodiversity. This would both satisfy the stated aims of the Biodiversity Convention and guarantee long-term food security for all.

Notes

1. See *Food for Our Future, Food and Biotechnology*, Food and Drink Federation, London, 1995.

2. "And still the children go hungry" Geoffrey Lean, *Independent on Sunday*, 10 November, p.12, 1996.

3. Hardy, 1994.

4. See "Food-population: Experts want to break wheat's yield barrier" A. Aslan, *Inter Press Service*, October 18, 1996.

6. Lester Brown of the World Watch Iinsitute, quoted in Goldsmith and Hildyard, 1991.

7. Goldsmith, 1992.

8. Hildyard, N. (1991). An open letter to Edouard Saouma, Director-General of the Food and Agriculture Organization of the United Nations. *The Ecologist* 21, 43-46.

9. Watkins, 1996.

10. " Seed action in Germany" E. Beringer, *Landmark* July/August, p.13.

11. Hildyard, 1996, p.282.

12. See DeAngelis, 1992; Pimm, 1991.

13. Moffat, 1996.

14. Raven, 1994.

15. Altieri, 1991.

16. "Throwing out the baby with the bathwater" C. Emerson, *On the Ground* September, p.2.

17. Shiva, 1993.

18. See Goldsmith, 1992; Shiva, 1993.

19. Meister and Mayer, 1994.

20. Mikkelsen *et al*, 1996; see also Ho and Tappeser, 1997.

21. Cox, 1995.

22. Perlas, 1994.

23. See note 15.

24. Shiva, 1993, Introduction; see also note 2.

25. *Alternative Agriculture, Report of the National Academy of Sciences*, Washington,DC, 1989.

26. Kothari, 1994.

27. "Seed action in Brazil" *Landmark* July/August 1996, p.10.

28. See note 2.

29. See *Food for Our Future, Food and Biotechnology*, Food and Drink Federation, London, 1995, p.5.

30. Shiva, 1993.

31. Nordlee *et al*, 1996.

32. Frank and Keller, 1995.

33. Lemke and Taylor, 1994.

34. Inose and Murata 1995.

35. See Ho and Steinbrecher, 1997, for a detailed critique of the 1996 FAO/WHO Food Safety Report, especially of the principle of 'substantial equivalence', which amounts to 'don't need - don't look - don't see'. See also Levidow *et al*, 1996. For up-to-date articles on GM foods, see Jaan Suurkula's home page: <http://home1.swipnet.se/~w-19482/indexeng.htm>

36. See Jorgensen and Anderson, 1994; Mikkelsen *et al*, 1996.

37. See Eber *et al*, 1994; Darmency, 1994.

38. Eber *et al*, 1994.

39. Rissler and Mellon, 1993.

40. See Holmes and Ingham, 1995.

41. See Cooking, 1989.

42. Reported by Perlas, 1995.

43. Wahl, *et al*, 1984.

44. See Doerfler, 1992; also Chapter 10, this volume.

45. Finnegan and McELroy 1994.

46. See Colman, 1996; Lee *et al*, 1995, and references therein.

47. "Monsanto swallows Calgene whole", Vicki Brower, *Nature Biotechnology* 15, 213, 1997.

48. See "Pests eat Monsanto's profits", *GenEthics News* issue 13, p.1, 1996, also "Bt cotton fiascos in the US and Australia", *Biotechnology Working Group: Briefing Paper Number 2*, BSWG, Montreal, Canada, May 1997.

49. Reynolds *et al*, 1996.

50. Johnston, 1989.

51. Johnston, 1989.

52. Hyrien and Buttin, 1986.

53. See most recent review by Estruch *et al*, 1997.

54. "Insecticide preservation policy: to be or not Bt", Jeffrey L. Fox, *Nature Biotechnology* 14, 687-688, 1996.

55. Estruch *et al*, 1997.

56. Bauer, 1995.

57. "UK's on-off affair with Ciba-Geigy's Supermaize", *The Splice of Life*, vol. 3, Issue 1, pp. 5-6, 1996.

58. Smirnov *et al*, 1994. I thank Jaan Suurkula for this information.

59. Schluter *et al*, 1995.

60. Green and Allison, 1994.

61. Allison, 1995.

62. Paulkaitis and Rossinck, 1996.

63. Cummins, 1994.

64. Heitman and Lopes-Pila, 1993.

65. Cummins, 1997. I thank the author for sending this article to me.

66. Schubbert, *et al*, 1994.

67. Cited in "Can DNA in food find its way into cells?", Philip Cohen, *New Scientist*, 4 January, p. 14, 1997.

68. Courvain *et al*, 1995, p.1207.

Chapter 10

The Immortal Microbe and the Promiscuous Genes

There can be little doubt remaining that horizontal gene transfer is responsible for both the emergence of new and old pathogens and multiple antibiotic resistances. The escalation in the emergence of pathogens and antibiotic resistances over the past decade coincides with the commercialisation of genetic engineering biotechnology. Many pathogens have crossed.species barriers in acquiring genes from different kingdoms that are involved in their ability to cause diseases. Genetic engineering is inherently hazardous because it depends precisely on designing gene transfer vectors to cross wide species barriers. Genetic engineers are motivated by the sole ambition to design more and more aggressive and wide host-range gene transfer vectors, and are quite unable to see the hazards involved. We are already experiencing a prelude to the nightmare scenario of uncontrollable, cross-species epidemics that are invulnerable to treatment. It is time we called a halt, to reconsider the strategy for the future of humanity.

Microbes as Friend and Foe

To understand why genetic engineering biotechnology is so inherently hazardous, we have to appreciate the prodigious power of microbes to proliferate and the protean promiscuity of the genes they carry, with their ability to jump, to spread, to mutate and recombine.

Microbes are ubiquitous. They live in abundance in the soil, in the terrestrial and aquatic environments, in the air we breathe, on our skin and in our bodies. Most of the time, they have a benign, balanced relationship with us, so that they do us no harm and, in many cases, a lot of good. Jim Lovelock and Lynn Margulis,[1] proponents of the Gaia hypothesis - the idea that the entire earth is a self-regulating system - show how different microbes are key players in the regulatory mechanisms that maintain the conditions on earth suitable for all life forms. It is well-known that microorganisms in the soil are indispensable for recycling nutrients for the growth of crop plants, and that bacteria in the gut of a healthy person can provide vitamins and cofactors, and aid digestion. Humans, indeed, have used microbes for centuries to make beer and wine, bread, yoghurt, cheese, sausages, miso, soy sauce and many other products, without experiencing harm. However, when that balanced ecology is disturbed, bacteria can turn virulent and cause debilitating, lethal diseases. And when we wage war on them with a succession of more potent antibiotics, they counter with ever more sophisticated antibiotic resistances

While mammalian cells, like our own, typically take a day or more to double in number, bacteria take only tens of minutes. A millilitre of tightly-packed *E. coli* contains a million-million cells, enough to infect the entire population of the world. A single bacterium will take less than a day, or 40 doublings, to multiply to that number when the conditions are right. On the other hand, it can persist indefinitely in a quiescent state, without multiplying, but mutating, or acquiring new genes, so that, given the opportunity, it will once again proliferate. They are, to all intent and purposes, immortal.

The genes these microbes carry have an even more tenacious hold on life. They have a greater propensity to proliferate, to spread promiscuously to different microbes as well as 'higher organisms', giving them the opportunity to multiply, mutate and recombine into new variants, to fully realize their protean potential, to do great harm, or not, in response to different ecological conditions.

The problem we face is that commercial-scale genetic engineering together with the profligate use of antibiotics in intensive farming and medical practices are creating just the sort of conditions for the microbes to do the greatest harm.

The Case of Antibiotic Resistance

Antibiotics were first introduced in 1944 for the treatment of microbial infections. Microbial geneticists in those days did not foresee the rapidity with which antibiotic resistance would evolve, based on the very low spontaneous mutation rates observed.[2] Antibiotics became widely used to control infectious diseases. But the euphoria about them was short-lived, as microbes soon developed resistance to antibiotics. By the early 1980s, fewer than 10% of all cases of *Staphylococcus* infection responded to treatment with penicillin, compared with almost 100% in 1952. Resistance to antibiotics has become widespread, with new resistances arising as soon as novel antibiotics come into use, and outbreaks of resistant bacteria now commonplace in hospitals. By 1990, nearly all common pathogenic bacterial species had developed varying degrees of antibiotic resistance. These include, besides *Staphylococcus aureus* (toxic shock syndrome, post-operative infections), *Streptococcus aureus* (toxic shock-like syndrome) *S. pneumoniae* (pneumonia) *S. pyogenes* (rheumatic fever) *Haemophilus influenzae* (meningitis) *Mycobacterium leprae* (leprosy), *Neisseria gonorrhoea* (gonorrhoea), *Shigella dysenteriae* (dysentery) and several other species of microbes that infect the human gut: *E. coli, Klebsiella, Proteus, Salmonella, Serratia marcescens, Pseudomonas, Enterococcus faecium, Enterobacteriaceae* and *Vibrio cholerae* (cholera). Multiple antibiotic resistance has also emerged. An Australian research team treated a patient infected with a strain of *Staphylococcus* that was resistant to 31 drugs including cadmium, penicillin, kanamycin, neomycin, streptomycin, tetracycline and trimethyloprim.[3] In a series of investigations, they showed that the various resistance capabilities were due to genes carried on different plasmids that could be separately passed on from one bacterium to another, most frequently by means of conjugation.

Soon after antibiotics were introduced in medicine, it was discovered that livestock lived longer when they were fed antibiotics to overcome infections. This led to prophylactic treatments, so that antibiotics were routinely fed to chickens, cattle, pigs and dairy cows. This also extended the shelf-life of meat, poultry, eggs and dairy products. In the 1970s it was discovered that giving chickens high doses of antibiotics resulted in the appearance of resistant *Salmonella* strains in both the meat and the eggs. By 1983, mutant strains of antibiotic-resistant *Salmonella* that attacked human beings had emerged.[4]

One of the clearest and most disturbing examples of mutant bacteria crossing from domestic animals to human beings, and of the transformation of a previously benign strain into a pathogen is *Escherichia coli*, a common bacterium inhabiting the intestine of all human beings and many other mammalian species. Most of the time, *E. coli* is harmless, which is why it is the darling of genetic engineers, who have used it and its plasmids routinely to clone genes since the 1970s. *E. coli* is the most manipulated organism, and genes from species in practically every other kingdom have been transferred to and cloned in it. Perhaps it is not surprising that E. coli has emerged as a major pathogen.

In 1982, however, a new strain, *E. coli* 0157: H7 appeared which, far from being benign, caused dangerous haemorrhages of the colon, bowel and kidneys in human beings. It broke out suddenly in several states in the USA.[5] Since then, many outbreaks have occurred all over the world, and in increasing frequencies. A mass outbreak in Japan in 1996 affected 9,000, with 12 deaths in children, and a further outbreak occurred in 1997. In Scotland, a series of outbreaks in 1997 claimed 20 lives and made hundreds ill.

DNA fingerprinting (see Chapter 12) backs up epidemiological evidence indicating that *E. coli* 0157:H7 arose recently and, up to 1993, was restricted to cattle in the USA, Canada and Great Britain but absent in cattle in the Far East (Thailand). *E. coli* mutates fairly rapidly. Hence, the low degree of genetic variation observed among the 0157:H7 isolates up to 1993 suggests that they originated from a single clone, which may have since spread to the Far East (Japan).

Most cases came from contaminated meat, but the mass outbreaks in Japan were traced to white radishes, which may have come from contaminated soil. The strain involved is moderately resistant to ampicillin and tetracycline and appears also to have acquired the ability to produce Shigella-like toxins, probably by horizontal gene transfer from *Shigella*.[8]

Studies of dozens of emergent species showed that genes for antibiotic resistance and virulence often resided in the same regions of the bacteria's DNA. On account of the misuse of antibiotics in both agriculture and medicine, *E. coli* strains were rapidly acquiring broad ranges of antibiotic resistance during the 1970s and 1980s. By the 1990s, super-strains resistant to multiple antibiotics were isolated. Two strains of *E. coli* appeared in a hospital transplant ward outside Cambridge that were resistant to an astonishing range of antibiotics: imipenen, as well as cefotaxime, ceftazdime, ciprofloxacin, gentamicin, ampicillin, azlocillin, coamoxiclav, timentin, cephalexin cefuroxime, cefotaxime, cefamandole, streptomycin, neomycin, kanamycin, tobramycin, trimethoprim, sulfamethoxazole, chloramphenicol and nitrofurantoin.[7] Only one commonly-used antibiotic remained effective: amikacin. If the strains ever became resistant to that drug, they would be totally invulnerable to treatment by antibiotics.

A Profusion of Mechanisms and Genes

A profusion of biochemical mechanisms and genes are involved in antibiotic resistances mechanisms , fully matching, if not surpassing, the variety of targets at which antibiotics are aimed.[8] Most antibiotics have a single primary target, usually a single step in the synthesis of macromolecules essential to the bacterial cell, and preferably do not affect their eukaryotic host cells. Thus, several classes of antibiotics are aimed at different steps in the synthesis of the bacterial cell wall. These include the penicillins or ß-lactams, vancomycin and ristocetin, D-cycloserine, fosfomycin and bacitracin. Other classes of antibiotics are aimed at bacterial enzymes involved in the synthesis of DNA, RNA and proteins, which differ from their eukaryotic counterparts, although their effects may not always be so selective, and are also accompanied by varying degrees of toxicity. Metronidazole, quinolones, nalidixic acid and novobiocin target DNA and DNA synthesis, while actinomycin and rifamycin inhibit transcription. Translation is targeted by a wide variety of antibiotics. Streptomycin and other aminoglycosides such as kanamycin, neomycin, gentamicin, tobramycn, aikacin and netilmicin; tetracyclines, puromycin, chloramphenicol, erythromycin, lincomycin and clindamycin, and fusidic acid. In addition, antimetabolites such as sulfonamides and trimethoprim inhibit the synthesis of essential metabolites.

Resistances to every class of antibiotics have been found. One of the commonest mechanisms of resistance is the degradation or modification of the antibiotic by specific enzymes. Thus, the beta-lactams are degraded by lactamases all of which have arisen from one ancestral gene. On the other hand, aminoglycosides are inactivated by specific modifying enzymes encoded by at least 30 different genes, while chloramphenicol is inactivated by many acetyltransferases encoded by at least a dozen genes. Mutation of specific antibiotic target is also frequently involved in resistances. For example, streptomycin is overcome in *Mycobacterium* by mutations in the ribosomal RNA (rRNA) molecules required for protein synthesis, so that they have reduced affinity for streptomycin. Antibiotics may be inactivated simply by binding permanently to specific proteins. A metabolic bypass may appear to a reaction inhibited by the antibiotic, or the inhibited protein may simply be overproduced by the bacteria in order to overcome the inhibition. Apart from those mechanisms which are specific for each antibiotic or chemical class of antibiotics, broad-range resistances are achieved by setting up permeability barriers to antibiotics entering the cell, or, more importantly, by increased rates of pumping them out of the cell. More than twenty different genes in 4 gene families have been identified in the latter category. They code for transmembrane proteins - active efflux pumps - that transport substances out of the cell using ATP or protons as sources of energy. Many confer multi-drug resistances, as they recognise a broad range of unrelated chemical substances. Quite often, several biochemical mechanisms are involved in giving protection against a single antibiotic, resulting in a very high level of resistance. Cross resistances are also common within a class of antibiotics such as the aminoglycosides (see Chapter 9).

Where do the genes conferring antibiotic resistances come from? And how do they manage to spread so quickly? Let us first address the question of the origin of the genes.

Origins of Antibiotic Resistance Genes

There are several sources of antibiotic resistance genes. The first comes from organisms producing the natural antibiotics, such as penicillin, which also produce the inactivating enzyme. As new ß-lactams are introduced, the genes encoding the ß-lactamases undergo mutational changes, often involving only one base substitution leading to a single amino acid difference in the enzyme, which would be sufficient to enable the enzyme to recognize the new drug and to break it down. In this way, the enzyme extends its spectrum of resistances to new drugs. Numerous mutational variants of ß-lactamase have already been identified, showing the ease with which appropriate mutational changes can arise in this gene. The variants derive from different genes which can all be traced to a single ancestral gene. The versatility of the beta-lactamase gene is such that when a series of inhibitors are made against the enzyme and used in combination with the ß-lactam antibiotic - a kind of belt and braces approach - mutant enzymes arise which not only break down the new antibiotic, but are, at the same time insensitive to the inhibitor.

More often, however, it is not at all clear where the antibiotic resistance genes come from. New genes seem to appear from nowhere. For example, the many genes involved in resistances to different aminoglycosides are not related to one another. Some are thought to have arisen from mutations of pre-existing genes encoding enzymes catalysing normal metabolic reactions in the cell. Others may involve activation of previously 'cryptic' or hidden genes in the bacterial genome. Or they may have been acquired by gene transfer from another bacterium. The active efflux mechanisms associated with multi-drug resistance in *E. coli* are believed to have originated from *operons* (multigene units under the control of a regulator gene) in free-living bacteria which enable the latter to respond to noxious substances in the environment by synthesising enzymes to break these substances down.

Before we go on to consider gene transfer in the spread of antibiotic resistance, I shall deal with the supposed natural selection of random mutations in the origin of antibiotic resistance genes.

The Irrelevance of 'Random' Mutations

To measure the rate of 'spontaneous' mutations to antibiotic resistance in a strain of bacteria, a small amount of a dilute suspension of the bacteria is spread on a nutrient agar-plate. This has been impregnated with a concentration of the antibiotic high enough to kill non-resistant bacteria, so that only those with a pre-existing mutation which happens to confer resistance will survive and form colonies. A colony appears as a round greyish-white spot on the agar plate, formed by the growth and multiplication of one original resistant bacterium. The spontaneous mutation rate for antibiotic resistance estimated in this way is typically 10^{-9} or less, and is about the same for all other genes in the genome. This is the rate of *random* mutations, in other words, mutations that have no direct correlation to the environment, or the selective regime. Based on these estimates, the early microbiologists did not believe that resistance to antibiotics would pose a problem when antibiotics first came into use. So why did antibiotic resistance evolve and spread so rapidly?

The neo-Darwinian explanation is that it is due to the intense 'selective pressures' exerted by the antibiotics, so that even extremely low rates of 'spontaneous' random mutations that happen to confer resistance to the antibiotic will be rapidly selected for, while those which do not happen to have the mutation will die out. This account is so far from the truth of what actually happens that I am appalled to read how many staunch neo-Darwinists are still at pains to defend it.

The evolution of antibiotic resistance is a paradigmatic example of the fluid genome which bacteria share with all organisms. In addition, because bacteria are potentially so good at exchanging genes, their fluid genomes are also getting rapidly de-localised. Genomes mutate, expand, contract and rearrange. Parts of them have semi-autonomous or autonomous existences and can travel between species that do not interbreed; in the process, they recombine and further mutate.

All bacteria, with the exception of those that live in hot springs, are killed by boiling or cooking, which is therefore the surest way to prevent infection. They can also be killed by high enough concentrations of antiseptics or antibiotics to which they are susceptible. However, at lower concentrations of the noxious agent, the bacterium may fail to grow at first but, after some time, will develop the required resistance to the drug which enables it to grow, multiply and form colonies. This ability to develop drug resistance, described in Chapter 8, is common to cells of all species - bacteria, fungi, plant and animal, without exception. It is a physiological response shared by *all* the cells or organisms of the population, and is not due to selection of pre-existing random mutations. It is essentially similar to the 'directed mutations' or 'adaptive mutations', also described in Chapter 8, which involve a wide diversity of mechanisms. To insist on interpreting them as the result of selection of random mutations is simply a desperate attempt to salvage the neo-Darwinian dogma that continues to serve the existing orthodoxy, as well as the biotech interests.

During a discussion on one of the papers presented at a conference on antibiotic resistance, precisely this question was raised - as to whether antibiotic resistance in bacteria is related to multi-drug resistance, which regularly arises in cancer cells in the course of chemotherapy. The answer, provided by one of the participants, S.B. Levy of Tufts University School of Medicine, was as follows,

"We have described an analogous multi-drug resistance system in *E. coli.* Chromosome-mediated multi-resistance emerges upon the use of a single drug, tetracycline or chloramphenicol. The mechanisms of resistance to the different antibiotics are very different. A similar phenomenon has been described among parasites. One must distinguish between extrachromosomal resistance elements, e.g. plasmids and transposons, and chromosomal adaptations to environmental stress. Mammalian cancer cells, bacteria and parasites all seem to be able to respond to toxic external agents by turning on cryptic chromosomal genes that aid their survival."[9]

Levy was distinguishing between endogenous mechanisms that turn on cryptic chromosomal genes and genes acquired by horizontal gene transfer through extrachromosomal elements or plasmids. We shall see later that this distinction is no longer useful, as genes can move easily between chromosomes and plasmids.

Horizontal Gene Transfer - the Genetic Meltdown

The other main reason that neo-Darwinism got it wrong is because competition is irrelevant among bacteria: they share their most valuable assets for survival - genes encoding resistance mechanisms against antibiotics. Horizontal gene transfer has emerged as a major mechanism for the spread of antibiotic resistance only since the late 1980s when a Conference on Antibiotic Resistance was co-sponsored by the Environment Protection Agency and the National Science Foundation of the USA[13] to look into the possible mechanisms and factors affecting such gene transfer. The first definitive evidence for horizontal gene transfer came from DNA sequence analysis of the genes for neomycin-kanamyin resistance from *Staphylococcus aureus*, *Streptococci* and *Campylobacter sp.*[14] which were found to be essentially identical. Further evidence of horizontal gene transfer, by DNA sequencing or other methods of characterising DNA, were subsequently obtained for a host of other antibiotic resistances.[14]

There are three mechanisms for gene transfer in bacteria: by direct uptake of DNA (*transformation*) through genes being carried by viruses that infect bacterial cells (*transduction*) and through a mating process requiring cell to cell contact (*conjugation*). These processes have been studied most extensively in *E. coli*, although it is becoming clear that what is true for *E. coli* does not necessarily apply to other species. The scope of our ignorance is enormous but enough is already known to give serious cause for concern.

Transformation frequencies are high

Transformation is the first mechanism for artificial gene transfer used in the experiment which proves DNA is the transforming genetic material (see Chapter 7). Transformation among microbial populations in the environment has been extensively reviewed recently,[16] showing that it is extremely widespread. The process involves release of DNA into the environment, binding of the DNA to the surface of bacteria, entry into cell and recombination with bacterial chromosome or, in the case of plasmids, it can involve reconstitution of the plasmid. All these stages are facilitated by bacterial proteins. Both chromosomal and plasmid DNA are able to transform bacteria, the frequency of transformation varying with the physiological state of the cell, the presence of salts, and the participation of diffusible factors excreted from the bacteria. The nature of the DNA, its source, size and state, may also be important. Generally, transformation is most frequent within the same species, as its successful integration into the recipient genome depends on the degree of homology, i.e. similarity in base sequence, between the transforming DNA and the host DNA. It also depends on the DNA being resistant to the restriction enzymes in the host cell which break down foreign DNA. However, homology is not necessary for successful transformation to occur. Cross-species, cross-genera and even cross-order transfers have been observed with chromosomal DNA. Plasmid DNA, in particular, have effected cross-kingdom transformations among Eubacteria, Proteobacteria and Cyanobacteria.

It is of interest that DNA is not only released into the environment by the death of cells, but is actively excreted by living cells during growth. Some species export DNA wrapped in membrane-derived vesicles. The DNA in a culture slime can be more than 40% of the dry weight. Thus, the environment is extremely rich in DNA. Marine water contains between 0.2 and 44μg per litre, fresh water contains between 0.5 and 7.8μg per litre; while freshwater sediment has a high concentration of 1μg per gram. Although enzymes breaking down DNA (deoxyribonucleases, DNases) are found in the environment, DNA is protected from degradation by adsorbing to detritis, humic acid and, in particular, clay and sand particles. Adsorbed DNA is equally efficient in transforming cells. Thus, the half-lives of DNA in soil is 9.1 hours for loamy sand soil, 15.1 h for silty clay soil and 28.2 h for clay soil. While half-lives in waste water are typically fractions of an hour, those in freshwater and marine water are 3 to 5 hours, with a high of 45 to 83 h on the ocean surface and extremely high values of 140 and 235 hours for the marine sediment. Adsorption of DNA to solid particles is a very rapid process, which means that DNA released into the environment can survive indefinitely and maintain its potential to transform other species. Transformation may be a major route of horizontal gene transfer: frequencies obtained under different environmental conditions, using artificial vectors, are found to be generally quite high, ranging between 10^{-2} to 10^{-5} per recipient cell. A special form of bi-directional transformation by cell contact and fusion is now known to be widespread (reviewed by Yin and Stotzky).

Transduction is substantial in aquatic environments

Transduction takes place by means of *bacteriophages* (viruses that infect bacteria) which are usually host specific, and are hence not expected to transfer genes between unrelated species. They consist of a viral DNA wrapped up in a protein coat which binds to specific receptor sites on the bacteria. Once bound to the receptor sites, the bacteriophage injects its DNA into the bacterium, where either it directs the synthesis of many bacteriophages which break open (lyse) the cell, or the phage DNA may insert itself into the bacterial chromosome. The bacteriophage can transfer genes by mis-packaging a piece of bacterial DNA in the viral protein coat, or by packaging a viral DNA that has a piece of bacterial DNA stuck to it. When the virus infects a second cell, the bacterial DNA is transferred and may then become integrated into the new bacterial chromosome. Although transduction is normally limited by host range, it may serve as a vehicle for transposable elements that can establish themselves in an otherwise unrelated host and hence contribute to cross-species, cross-genera gene transfer. Also, bacteriophages themselves evolve by horizontal gene transfer and recombination, leading to broadening of their host ranges. One study[17] has shown that there may be as many as 10^8 bacteriophages per millilitre in aquatic environments, so that one-third of the total bacterial populations is subjected to a phage attack every 24 hours, each capable of transducing genes from another species.

Conjugation is promiscuous

Conjugation was previously thought to be a species-specific process requiring complex, complementary interactions between the genetic 'donor' and the genetic 'recipient'. Conjugation depends on certain conjugative plasmids, extrachromosomal pieces of DNA that are replicated independently of the chromosomal. They carry genes required for the process of conjugation. These plasmids fall into different *incompatibility groups*, defining mating relationships that result from distinct systems of DNA replication where both donor and recipient genes are involved. According to a review in 1993,[18] there are seven or more incompatibility groups of over a hundred known plasmids, *most of them carrying one or more antibiotic resistance genes*. Plasmids in the same incompatibility group cannot be maintained in the same bacterial cell at the same time. However, there are common mechanisms for gene transfer that are interchangeable. One of these codes for the *pilus*, a tube-like structure connecting the donor with recipient cells in conjugation, through which plasmid DNA passes from the donor to the recipient. In other cases, variation between products of genes are specific and not interchangeable, such as those involved in the transfer of the plasmid. Several incompatibility groups of plasmids are 'promiscuous' in that they have a very broad host range.

Conjugative plasmids can integrate themselves into the bacterial chromosome. In such cases, conjugation will take place, and the integrated plasmid will also drag the bacterial chromosome with it across the conjugation tube from donor to recipient, leading to transfer of bacterial DNA as well as plasmid DNA. Donor DNA will then recombine with recipient DNA to generate new genetic recombinations. The DNA is thought to be passed in a single stranded form, which is resistant to breakdown by enzymes attacking double-stranded DNA (nucleases and restriction enzymes), thus ensuring a high success rate of gene transfer. The single-stranded DNA then directs the synthesis of the complementary strand to restore double-stranded DNA in both donor and recipient.

Recent studies indicate that conjugation has an extraordinarily wide host range, involving diverse, complex mechanisms that are not yet well understood.[19] The 'promiscuous' plasmids overcome species barriers, effecting horizontal gene transfer between phylogenetically unrelated species. In addition conjugative transposons (mobile genetic elements) have been discovered, which mediate their own transfer to recipient cells during conjugation and insert into the recipient chromosome. They can also jump from one site to another on the same chromosome, or from chromosome into plasmids and *vice versa*. A special class of such conjugative transposons are 'integrons',[21] which support the site-specific integration of antibiotic-resistant gene 'cassettes' within the integron, so that each 'cassette' is then provided with ready-made signals for expression. Each integron can carry several cassettes, each encoding a different antibiotic resistance. It can facilitate recombination between different cassettes to form exotic gene-fusions that code for multi-functional proteins. Integrons can also jump from the bacterial chromosome into a plasmid and become transferred to another bacterium during conjugation. During conjugation, genetic material goes not only from donor to recipient, but also in reverse. Such retro-transfers of genetic material have only recently been discovered.

The result is that there could be very few barriers remaining to horizontal gene transfer. Antibiotic resistance genes, especially those carried on plasmids and transposons, can, in principle, cross species as well as genera and even kingdoms, as we shall see.

Tetracycline-resistant genes, for example, are now found to be shared between many genera, one of them, the *TetL* gene, is found in *Actinomyces, Clostridium, Enterococcus, Listera, Peptostreptococcus, Streptococcus, Staphylococcus and Bacillus.* Another gene, the *TetK*, is found in *Bacillus, Clostridium, Enterocococcus, Eubacterium, Listeria, Peptostreptococcus, Staphylococcus and Streptococcus.*[22] More recently, both of these genes have been found for the first time in soil bacterial species belonging to the genera *Mycobacterium* and *Streptomyces*, which can cause soft-tissue and skin infections.[23] Antibiotic- producing *Streptomyces* are believed to be the ancestral source of many aminoglycoside antibiotic resistance genes, but they are now regaining more of these genes from those other species. These findings are of particular significance, as they "suggest the potential for the spread of an antibiotic resistance gene into all environmental mycobacteria, including *Mycobacterium leprae*."[24]

Even chromosomally encoded penicillin-resistance genes (penicillin-binding proteins, pbps, required for the uptake of penicillin) are readily transferred. Transfers of *pbp* genes were found between *Streptococcus pneumoniae* and *Neisseria gonorrhoea* which subsequently recombined to generate new hybrid genes.[25]

The Emergence of New and Old Virulent Strains

The very same genetic mechanisms for horizontal gene transfer have been shown to be involved in the emergence of virulence among old and new pathogens since the mid-1980s. For example, there has been an increase since then in sporadic cases of very severe invasive *Streptococcus pyogenes* infections in Europe, North America and elsewhere. In an analysis of 108 isolates from patients in the USA with Streptococcal toxic-shock syndrome,[26] a toxin is frequently found encoded by a gene, *SpeA*, belonging to a bacteriophage which has become inserted into the bacterial genome. This gene was spread horizontally among divergent strains of *S. pyogenes*. In addition, genes encoding membrane surface proteins, which determine binding properties and virulence of the bacteria, have undergone numerous recombination events subsequent to horizontal gene transfer, giving rise to mosaic structures of the genes. These are superimposed on the accumulation of point mutations acquired as the result of interactions of the pathogens with the immunological system of the human host, which make antibodies against the membrane surface proteins of the bacteria. Similar findings were produced by the analysis of group A streptococci, isolated from a cluster of cases of serious infections over a 3-month period in Tayside, Scotland, in 1993.[27] DNA sequence analysis of virulence factors suggests that they may have recently been acquired by horizontal gene transfer.

In the past, only *Vibrio cholerae* strains of the type O1 were known to cause epidemics, while non-O1 strains were associated with sporadic cases. But a recent epidemic in Asia was caused by a non-O1 strain, *Vibrio cholerae* O139. It turns out that the new strain is identical to an earlier pandemic strain, O1 EL Tor, except for proteins involved in the capsule and the O-antigen synthesis. This was due to the acquisition of DNA inserted into and replacing part of the O antigen gene cluster. This suggests that O139 arose by horizontal gene transfer from a non-O1 strain into a type O1 strain.[28]

Pathogenic mycoplasmas possess proteins called adhesins and related accessory proteins which are required for adhering to cells and subsequent disease development. *Mycoplasma genitalium*, implicated in urethritis, pneumonia, arthritis and AIDS progression, was found to encode one adhesin and 2 adhesin-related proteins that shared substantial sequence similarities, as well as genome organisation, with those of another species, *M. pneumoniae*.[29] These were attributed to horizontal gene transfer events between the two species.

Most disturbing of all is the recent discovery that at least 10 unrelated bacterial pathogens, causing diseases from bubonic plague to tree blight, share an entire set of genes for invading host cells, which have almost certainly spread by horizontal gene transfer.[30] More than 20 genes are involved in a system that secretes damaging proteins directly into host cells. They were first discovered in *Yersinia*, a genus of several species that cause human disease including bubonic plague and intestinal infections. These proteins prevent bacteria from being taken up and destroyed by white blood cells. The system differs in different bacterial species. In general, it disables the host cell and makes it more hospitable to the bacterium. In some species. such as *Shigella*, the secreted proteins induce non-immune cells to internalise the bacteria. In *E. coli*, these proteins allow the bacteria to stick to the intestinal epithelium. The disturbing evidence is that at least some of the proteins are typical of eukaryotic cells, and are targeted at eukaryotic biochemistry. This has led to the speculation that these genes were acquired from eukaryotic cells. *One way in which this might have happened is through genetic engineering biotechnology itself, in which arbitrary recombinations of gene sequences of infectious genetic elements are routinely made.*

The evidence is now overwhelming that horizontal gene transfer has been responsible for both the rapid spread of antibiotic resistances and for the emergence of virulent strains of pathogens. This raises the question of whether the full range of mechanisms for horizontal gene transfer has always existed. This is almost impossible to answer, as it is only through the application of recently developed genetic engineering techniques that the gene transfer vectors and mechanisms were discovered.

A study carried out in 1983 on strains of bacteria isolated between 1917 and 1954, which is well into the antibiotics era, showed that none of the strains carried any antibiotic resistance.[31] However, 24% of them encoded genetic information for the transfer of DNA from one bacterium to another; and that, from at least 19% of the strains, conjugative plasmids carrying no antibiotic resistance were transferred to the laboratory strain *E. coli* K12. Thus, while conjugative plasmids do pre-date the antibiotics era, their prevalence and the range of drugs to which they confer resistance have greatly increased since.

Horizontal Gene Transfer is the Greatest Threat to Public Health

I do not think it is an exaggeration to say that horizontal gene transfer (and consequent recombination) is the greatest threat to public health facing us to-day, especially if commercial-scale genetic engineering biotechnology is allowed to continue unchecked. I must emphasise again that there has been a dramatic recent escalation in the evolution of antibiotic resistance and in the incidence of virulent infections. For example, some countries in Europe have witnessed a staggering 20-fold increase in incidence of *Salmonella* infections over the past 10 to 15 years. *Salmonella* infections are contracted mainly through the consumption of raw or undercooked contaminated meat, poultry, eggs and milk. And, since the beginning of the 1990s, multiple antibiotic resistances have arisen in *S. typhimurium* which, together with *S. enteritidis*, account for the majority of the epidemics.[32]

The dramatic increase in virulent infections and antibiotic resistance within the past 10 to 15 years is usually put down to the profligate use of antibiotics in intensive farming and in medicine. Antibiotics are hazardous, not only because they provoke antibiotic resistances to evolve, but because the presence of antibiotics can actually increase the frequency of horizontal gene transfer ten to 100-fold.[33] In one experiment, the antibiotic oxytetracycline increased the frequency of transfer of a plasmid encoding multiple antibiotic resistance by 3 to 4 orders of magnitude.[34] It is known that ancestral antibiotic resistance genes have originated from antibiotic-producing microorganisms themselves. The possibility has been suggested, therefore, that what we call antibiotics may actually be sex hormones for the bacteria, enhancing conjugation between cells; and, furthermore, that the process is naturally regulated by the product of the 'antibiotic resistance' gene which shuts off the signal. Thus, the use of antibiotics not only leads to mutational transformations of antibiotic resistance genes, it enhances the spread of those very genes among pathogens.

The other factor we have to take into account is that the dramatic increase in virulent infections and antibiotic resistance has also taken place since commercial-scale genetic engineering biotechnology began. The spread of virulence genes and the generation of new pathogenic strains of bacteria and viruses cannot be attributed to the overuse and abuse of antibiotics *per se*. They may well be the consequence of other practices, such as the extensive genetic manipulations of microbial and other genomes that are throwing the ecology of microbes seriously out of balance. Although there is no direct evidence linking genetic engineering biotechnology to the spread of virulence and antibiotic resistance, there is clear evidence that horizontal gene transfer *is* responsible for both. And, there is no escaping the fact that the *raison d'etre* and aspiration of genetic engineering *is* to increase the facility of horizontal gene transfer, so as to create ever more exotic transgenic organisms.

Artificial gene transfer vectors make it much more likely to occur

It is not easy to transfer genes naturally between species. That is why such events were relatively rare in our evolutionary past. For example, analyses of 145 non-vertebrate globin gene sequences showed that there were probably two cases of horizontal gene transfer, one from the common ancestor of ciliates and the green algae to the ancestor of cyanobacteria, and the other from the ancestor of the yeasts to the ancestor of bacteria.[35] Natural gene transfer vectors - viruses, plasmids and transposons - have probably always existed, especially in the microbial species; but they are, to varying degrees, host specific, so that the frequency of conjugal transfers is higher between the same species than with other species.[36] Furthermore, there are endogenous cellular mechanisms that break down foreign DNA, fail to replicate it, and excise or inactivate foreign genes that do get inserted into the genome (see Chapter 8).[37] These mechanisms are also responsible for the instability and silencing of transferred genes in transgenic organisms, which is posing a problem for the technology (see Chapter 9).[38]

Genetic engineering biotechnology is based on constructing a wide range of aggressive hybrid vectors designed both to overcome species barriers and, increasingly, to overcome cellular mechanisms that break down or inactivate foreign DNA.[39] Artificial gene transfer vectors are made by joining together bits of natural gene transfer vectors - viruses, plasmids and transposons - from different sources, so they are essentially promiscuous. And, indeed, 'naturally occurring' promiscuous vectors have been used in genetic manipulations. These artificially constructed vectors effectively smuggle genes into cells that would otherwise exclude them, and enable the genes to replicate in cells that would otherwise not recognise them or break them down. In addition, because the vectors already possess sequence homologies to a wide range of disease-causing viruses, they can recombine with them most easily to generate new, broad host-range viruses. As mentioned in Chapter 1, these viruses are already with us today. Although artificial gene transfer vectors are supposed to be crippled and, hence, not mobilisable by themselves, they can easily be helped by other plasmids and viruses that are present in all organisms. It is significant that, although there is abundant evidence of horizontal gene transfer from sequence analysis of clinical isolates, most laboratory demonstrations of horizontal gene transfer have been carried out using artificially constructed gene transfer vectors,[40] which clearly show that they can mediate unintended cross-genera and cross-kingdom gene transfers. Naturally isolated vectors - even those that carry antibiotic resistance genes and have therefore already undergone evolution by horizontal gene transfer - have not exhibited the same wide range of transfer capabilities. The Ti (tumour-inducing) plasmid of *Agrobacterium*, which is said to mediate conjugation between *Agrobacterium* and plant cells,[41] effecting a cross-kingdom gene transfer, actually does so in a genetically manipulated form. This *Agrobacterium* Ti plasmid is widely used to make gene transfer vectors for genetic engineering plants. Most, if not all, transgenic plants currently released or awaiting release have been constructed with this kind of vector. The potential for Ti plasmid-mediated secondary horizontal gene transfer from transgenic plants is unlimited. Yet, *no monitoring for horizontal gene transfer is required under current regulations.*

As mentioned in Chapter 8, the microbial populations in the environment serve as a gene transfer highway and reservoir, enabling genes to be replicated, recombined, to spread from non-pathogens to pathogens, and to infect all other organisms. The release of genetically engineered microorganisms is especially hazardous.

Horizontal gene transfers have been directly demonstrated between bacteria in the marine environment,[42] in the freshwater environment,[43] and in the soil.[44] It is significant that, in all the experiments, horizontal gene transfers are mediated by specially constructed hybrid plasmid vectors, of the sort used in genetic engineering. Horizontal gene transfer occurs preferentially in interfaces between air and water and in the sediment, and especially under nutrient depletion conditions,[45] thus refuting the claim that nutrient-rich media are necessary to support horizontal gene transfer. Horizontal gene transfer has even been demonstrated in waste-water treatment ponds, the effluent from which is increasingly being used for irrigation in developing countries.[46]

Horizontal gene transfer is not limited to the external environment. It has been demonstrated between gut bacteria in mice and chickens,[47] and in the gut[48] as well as uro-genital and respiratory tract of human beings.[49]

Stephenson and Warnes wrote in 1995,

"The threat of horizontal gene transfer from recombinant organisms to indigenous ones is...very real and mechanisms exist whereby, at least theoretically, any genetically engineered trait can be transferred to any prokaryotic organism and many eukayotic ones."[50]

A year later, another molecular geneticist, Harding, who works on transgenic plants, admitted that, "...the potential for horizontal [gene] transfer may be greater than thought previously."[51]

While some geneticists recommend caution in the face of the already existing evidence for horizontal gene transfer, those still dominated by the genetic engineering mindset see the potential for horizontal gene transfer as new opportunities to be further exploited, dismissing people who present the evidence as expressing "scenario-based concerns of lesser validity!"[52]

What the public is up against is a selective blindness to evidence among the genetic engineers and a single-minded commitment to look solely for the exploitable - that is the hallmark of a bad science. To make things worse, the public is getting little or no protection from existing regulations (see below).

Genes carried by vectors can persist indefinitely in the environment

Genes carried by vectors can survive indefinitely in the environment, within thriving or dormant bacteria, or as naked DNA adsorbed to solid particles, where they are efficiently taken up by other microbes, as already described above.[53] The survival of crippled laboratory strains of bacteria and the persistence of DNA in the environment were first brought to our attention by Dr. Beatrix Tappeser of the Ecological Institute of Freiburg, who has been campaigning for many years in Germany and elsewhere in Europe against the inadequacy of regulation on genetic engineering biotechnology. Beatrix was another regular member of the group working with the TWN. She has a quiet, understated and business-like style of delivery which can be very effective. It was in this manner that she presented what must have been the most significant paper in the first of the TWN seminars. That was in New York in 1995, which was when the dangers of horizontal gene transfer first occurred to me.

According to a recent study in Eastern Germany, streptothricin was administered to pigs, beginning in 1982. By 1983, plasmids encoding streptothricin resistance was found in the pig gut bacteria. This had spread to the gut bacteria of farm workers and their family members by 1984, and to the general public and pathological strains of bacteria the following year. The antibiotic was withdrawn in 1990. Yet the prevalence of the resistance plasmid had remained high when monitored in 1993, confirming the ability of microbial populations to serve as stable reservoirs for replication, recombination and horizontal gene transfer,[54] in the absence of selective pressure. In a direct test of persistence of streptomycin-resistance,[55] researchers cultured many independent lines of a streptomycin-resistant mutant of *E. coli* in the absence of the antibiotic. They found that all retained the resistance after 180 generations. Furthermore, the lines had also, in the meantime, accumulated compensatory mutations in other parts of the genome that increased their competitive ability relative to the wild-type. Again, this demonstrates the irrelevance of the neo-Darwinian explanation with regard to the evolution of antibiotic resistance.

Bacteria and viruses can, indeed, apparently disappear as they go dormant, and then reappear in a more competitive form. This has been documented for a laboratory strain of *E. coli* K12 which, when introduced into sewage, went dormant and undetectable for 12 days before reappearing, having acquired a new plasmid for multi-drug resistance that enabled it to compete with the naturally occurring bacteria.[56] Dormant forms of bacteria and viruses can survive indefinitely as biofilms made up of extracellular matrices containing multiple species communities. These are organised assemblages which can be found in the circulatory system of our bodies as well as in the environment.[57] This is an active, albeit non-proliferative, mode of existence which bears a lot of resemblance to the cells of a multi-cellular organism (see Chapter 14). In this state, the bacteria continue to accumulate new mutations, to exchange genes and to come back with a vengeance when the ecological conditions are ripe.

These findings take on additional significance in the light of a recent report that the chemical treatments in waste-tanks of commercial aircraft are insufficient to inactivate pathogens (see Chapter 1). Commercial-scale contained users of genetically modified microorganisms (GMMs) routinely release large amounts of wastes after chemical or physical 'inactivation'. How adequate are these inactivation measures? Even if inactivation is effective, we now know that the large amount of recombinant DNA released can still be readily transferred to other bacteria by direct uptake (i.e. transformation). There is an urgent need to re-assess the safety regulation of contained use, as inadequately inactivated pathogenic and other dangerous GMMs may already be being routinely discharged into the environment. One 'environmentally conscientious' biotech company has even turned its inactivated transgenic sludge into a fertiliser for crop-plants.[58] Far from contemplating such re-assessment, the UK Health and Safety Executive (HSE) has just drafted a document which would allow commercial and other contained users to release certain classes of live GMMs into the environment as liquid wastes, simply on notification, and without the need to monitor for the survival and subsequent evolution of the GMMs.[59] I was advised by Professor Guënther Stotzky, an expert on horizontal gene transfer, to fight it at the "highest level of government".

Conclusion

There can be little doubt remaining that horizontal gene transfer is responsible for the emergence both of new and old pathogens and multiple antibiotic resistances. The escalation in the emergence of pathogens and antibiotic resistances over the past decade coincides with the commercialisation of genetic engineering biotechnology. Many pathogens have crossed species barriers in acquiring genes from different kingdoms that are involved in their ability to cause diseases. Genetic engineering is inherently hazardous because it depends precisely on designing gene transfer vectors to cross wide species barriers. Genetic engineers are motivated by the sole ambition to design more and more aggressive and wide host-range gene transfer vectors, and are quite unable to see the hazards involved. This situation is likely to get totally out of control. We are already experiencing a prelude to the nightmare scenario of uncontrollable cross-species epidemics that are invulnerable to treatment. It is time we called a halt, to reconsider the strategy for the future of humanity.

Notes

1. See Lovelock, 1979; 1996 and references therein.

2. See Davies, 1994.

3. Udo and Grubb, 1990.

4. Garrett, 1995, presents a detailed history of the rise of antibiotic resistance in Chapter 13 of her excellent monograph.

5. Riley *et al*, 1983.

6. WHO press release 41, 21 May, 1997
See Knight, 1993

See Volk *et al*, pp. 362-370.

9. Brown, *et al*, 1993.

10. The account given here is abstracted mainly from Davies, 1994, Nikaido, 1994 and Spratt, 1994. See Volk et al, 1995, pp. 253-284, for a description of the classes of antibiotics and their mechanisms of action.

See Miller and Sulavik, 1996

Levy and Novick, 1986, p.79.

See Levy and Novick, 1986.

14. Trieu-Cuot *et al*, 1985.

15. See Sougakoff *et al*, 1987; Manavathu *et al*, 1988; Kell *et al*, 1993; Amabilecuevas and Chicurel, 1993; Coffey *et al*, 1995; Bootsma *et al*, 1996. Horizontal gene transfer is recently reviewed by Yin and Stotzky, 1997.

16. Lorenz and Wackernagel,1994; Yin and Stotzky, 1997.

17. Bergh *et al*, 1989.

18. Ippen-Ihler and Skurray, 1993.

19. See Clewell, 1993; Yin and Stotzky, 1997.

20. Franke and Clewell, 1981; Clewell and Flannagan, 1993.

21. Stokes and Hall, 1989; 1992; Collis *et al*, 1993.

22. Roberts, 1989; Roberts and Hillier, 1990; Speer, Shoemaker and Salyers, 1992.

23. Pang *et al*, 1994.

24. Pang *et al*, 1994, p.1411.

25. Spratt, 1988.

26. Kehoe *et al*, 1996.

27. Upton *et al*, 1996.

28. Prager *et al*, 1995, Reidl and Mekalanos, 1995; Bik *et al*, 1995.

29. Reddy, *et al*, 1995.

30. Barinaga, 1996.

31. Hughes and Datta, 1983.

32. WHO fact sheet No.139 January, 1997.

33. See Davies, 1994; Mazodier and Davies, 1991; Torres *et al*, 1991.

34. Sandaa and Enger, 1994.

35. See Moens *et al*, 1996.

36. Pang *et al*, 1994.

37. Doerfler 1991, 1992.

38. Finnegan and McElroy, 1994.

39. Höfle, 1994.

40. Mazodier and Davies, 1991.

41. Stachel *et al*, 1986; Kado, 1993.

42. Frischer *et al*, 1994; Lebaron *et al*, 1994; Sandaa and Enger, 1994.

43. Ripp *et al*, 1994.

44. Neilson *et al*, 1994.

45. Goodman, *et al*, 1994.

46. Merzrioui and Echab, 1995.

47. Doucet-Populaire, 1992; Guillot and Boucaud, 1992.

48. See Anderson, 1975; Freter, 1986.

49. Roberts, 1989.

50. Stephenson and Warnes, 1995, p. 5.

51. Harding, 1996.

52. Mazodier and Davies, 1991, p.148.

53. Reviewed extensively by Jager and Tappeser, 1995.

54. Tschäpe, 1994.

55. Schrag and Perrot, 1996.

56. Tschäpe, 1994.

57. Costerton *et al*, 1994; Lewis and Gattie, 1991.

58. According to the Novo Nordisk 1996 Environmental Report (kindly sent to me by their Environment Director), the Danish authority sets the following limits for untreated release of GMMs in terms of colony-forming units (roughly equivalent to live bacteria): waste water, 10,000/ml, air 10,000/ml, solid waste, 10,000/g. Novo Nordisk also recycles inactivated GMMs as fertilisers for crops, under the trade name of NovoGro.

56. I telephoned the UK Health and Safety Executive (HSE) in June, 1997, to obtain information on the safety regulation of commercial-scale contained use. The per-

son at the other end of the line said they only had a small leaflet, which he would send to me, but, that if I required more detailed information, I was to ring another number for the health inspectors. I rang that number for the next two weeks, and never once was the telephone answered, nor was there an answer-phone service to leave messages on. When the leaflet failed to arrive after a week, I telephoned the HSE again. Another person assured me that the normal time it took for the leaflet to get to anyone was ten days. However, if I needed it right away, I could pay £10, or £5 for delivery in the next two days. I protested that, as a public department, and for information of such vital importance to public health, they really ought to be much more prompt and to provide it free of charge. He replied that the Department received hundreds of enquiries every day, more than they could cope with, and that, in any case, the leaflet-provision service was privatised. What's more, the service he was providing - answering calls from members of the public - was also privatised, and if I needed to address my complaints, I must do it elsewhere, as he was not the person responsible.

When I finally received the information, I discovered that the HSE, on the recommendation of the Advisory Committee on Genetic Modification (ACGM), was circulating a document drafted by the ACGM (Draft Guidance on Certificate of Exemption No. 1) which would allow commercial and other contained users to release certain classes of live GMMs into the environment as liquid wastes on notification, without the need to monitor for the survival and subsequent evolution of the GMMs. I wrote to oppose the draft, spelling out the dangers and calling for a full reassessment of current safety regulations on contained use. I sent copies of the letter to several members of Parliament who asked questions of the Departments of the Environment and Transport. The official line was reaffirmed in their reply, which I again challenged, quoting Stotsky's comments to me in an e-mail message.

Since then, the HSE has admitted that I have raised some important issues and is commissioning an 'independent' critical review of the literature in 1998. However, no actual targeted experimental research will be supported.

Chapter 11

Hello 'Dolly' Down at the Animal 'Pharm'

Why would anyone want to clone a sheep or a cow, let alone a human being? No one except the genetic determinist who believes an organism is nothing more than the sum total of its genetic makeup. Dolly was not even a clone. Somatic cells accumulate systematic and non-systematic changes in genomic DNA during development, which accounts for the low success rate of this so-called 'cloning' technique. The experiment is misguided. It is not the best way to generate identical clones, but to generate monstrous failures of development. It is irresponsible and unethical to claim otherwise.

The cloning and 'pharming' of livestock, the creation of transgenic animals for xenotransplantation and to serve as animal models of human diseases, are all scientifically flawed and morally unjustifiable. They also carry inherent hazards in facilitating cross-species exchange and recombination of viral pathogens. These projects ought not to be allowed to continue without a full public review.

A Frankenstein beyond Reproach

On Sunday, February 23, 1997, Ian Wilmut, an embryologist working in the Roslin Institute just outside Edinburgh in Scotland, announced that they had succeeded in 'cloning' a sheep from a cell taken from the mammary gland of an adult. The clone, named 'Dolly', then seven-months-old, was said to be genetically identical to the adult from which the cell had been taken. Public reaction was swift. Did it mean, people asked, that this could be done in humans? Were we nearer to cloning human beings? Why was this research allowed to go on at all? And why was it only coming to public attention now, some 10 years after the work had begun?

The headlines for the next few days were sensational. "Galileo, Copernicus - and now Dolly!"[1]... "The spectre of a human clone"... "In the past few days, we have lived through a change in our condition as momentous as the Copernican revolution or the splitting of the atom"[2]... "Scientists 'able to create human clone' "[3].

President Clinton of the United States was quoted as saying that the cloning of Dolly raised "serious ethical questions, particularly with respect to the possible use of this technology to clone human embryos." He told a panel of bioethics experts to report back to him in 90 days on the ethical and legal implications of the Edinburgh work. (In June, President Clinton imposed a ban on human cloning for 5 years.) On Thursday of that week, Sir Ian Campbell, Chairperson of the UK Human Genetics Advisory Commission, announced that the Commission was meeting that very day to deliberate the implications of the new cloning science.

The story was front-page in the *New York Times*. It dominated television newscasts and gave rise to endless streams of articles and talk shows. By Wednesday, the share price for PPL Therapeutics, which carried out the work in collaboration with scientists at the Roslin Institute, had risen by more than a third, to increase its market value by £25m. We are to make no mistake as to what is driving the science. It is important to stress that the Roslin scientists own no shares in the company, and will not benefit *directly*. However, the cloning technology has been patented jointly by PPL and the Institute. So the Institute will certainly expect to benefit from royalties, if not from continued research contracts and grants.

At first, the scientists involved, including Wilmut himself, dismissed the whole subject of cloning humans as science fiction. The technique was very difficult, they said. They had manipulated nearly 300 embryos to get one success. The fact that it could be done in sheep did not mean it could be done in humans. Besides, they pointed out, there was no need, and it was illegal in Britain in any case.[4] Then, a few days later, Wilmut admitted that it could be done in humans, although the Director of the Roslin Institute, Professor Grahame Bulfield, insisted they would not allow cloning to be used in harmful ways, and especially not for work on humans. Instead, he emphasised that the breakthrough could, in the long-term lead, to "a myriad new ways" to help humans. Herds of transgenic animals could be farmed for proteins, blood and organs. Gene therapy could provide cures for fatal diseases.[5]

Ian Wilmut himself offered the prospect that a human embryo, produced by the same methods, could be used to treat cancer and other life-threatening diseases. Human embryos created like Dolly would then be grown until key cells could be extracted from the embryo and used to treat human diseases. During the work, the embryo would die. In other words, human embryos would be farmed, or 'pharmed' like the transgenic animals mentioned by the Director of the Institute. The horror of this thought is tempered only by Wilmut's re-affirmation that cloning a human would be "technically difficult and ethically unacceptable". However, it recently transpired that patents on the technology filed by the Institute would cover all 'animals', including human beings.[6]

There are laws against human cloning in some, but not all, countries: in the United Kingdom, Spain, Germany, Canada and Denmark, for example. In the U.S., although federal funds cannot be used for research on human embryos, the position for privately funded research is unclear. Joseph Rotblat, British physicist and Nobel laureate, called for the establishment of an international ethical committee.[7]

While Wilmut welcomed President Clinton's reaction and accepted the need for the issues raised to be considered by biologists and professors of ethics, he was unapologetic about the technique. He expressed irritation at the continuing "atmosphere of criticism" surrounding his success. "Here we have a remarkable achievement, a world first, and there are people who seem to make a living out of spreading angst, he said.

"You cannot blame the scientists for making those kind of discoveries. We are not Frankenstein-type people. If we hadn't made the breakthrough somebody else would; the technology is out there. It is now up to society to decide how it should be used and we welcome any discussion of these matters."[8]

These are significant words, not least because they reveal the scientist's unspoken assumption that he can do no wrong. He is, by implication, simply following a natural obligation for the "advancement of science", an aim above reproach. In fulfilling this noble obligation, there can be no question of any personal responsibility to decide whether he should.

There are many forces at work that converge towards this eventuality. The pure motive of the advancement of science may be only one among them, personal advancement and prestige, a strong other. And one must not underestimate the importance of financial backing from the pharmaceutical industry, eager to reap the rewards of a growing market in reproductive biotechnologies. A substantial amount of the financial support for the research actually came from the taxpayer, via the Ministry of Agriculture Fisheries and Food, as was revealed when the Ministry announced it was withdrawing the group's funding barely a week afterwards.[9] The ostensible reason for this was that they had already succeeded in what they had proposed to do with the grant from the Ministry: cloning sheep.

It is significant that not one among the luminaries invited to comment on Dolly in the UK within the first week of the discovery was a woman. Women have been conspicuously absent from the scene. The only allusion to women was Wilmut's revelation[10] that the world's first cloned animal was named after the singer, Dolly Parton, because the cell used to create her came from the "impressive mammaries" of the adult sheep.

This 'cloning' technique is the latest development in an accelerating trend in industrialised societies to wrest control of reproduction away from women into the hands of expert scientists and, ultimately, of faceless corporations that turn reproduction into services and commodities.[11] (I say 'cloning' deliberately because, as I shall reveal later, the technique does not actually result in a genetically identical clone.)

It all began with the Pill and other methods of contraception which are predominantly aimed at women. Although the Pill is generally seen as giving women more choice and control, it also puts the entire burden of responsibility for parenthood and otherwise on them, leaving men completely free and 'blameless'. That is why we live in a society that still stigmatises single mothers. Women are not in control when they take the Pill (and suffer all the side-effects besides) because their partners are automatically absolved from any responsibility. After the Pill came *in vitro* fertilisation and infertility treatments, sex determination of embryos, surrogate motherhood, germline gene replacement therapy, and now, 'cloning', a method that bypasses fertilisation altogether. It is the logical culmination of the instrumental, exploitative science, that treats nature as so many objects to be manipulated for the benefit of 'mankind'. So embryos, even human embryos, can be turned directly into commodities, or else into 'pharm' animals to produce proteins, cells or organs to order, for those who can afford to pay.

But who would want to clone a sheep, or a cow, let alone a human being? No one except the genetic determinist who believes an organism is nothing more than the sum total of its genetic make-up and, perhaps, those who believe it is their right to exploit cloned human beings or animals for spare body parts. It is indeed genetic determinism that inspires the act, that simultaneously validates and legitimises it and makes it so compelling, not only for the scientists concerned, but for a substantial sector of the public who have become hooked on the genetic determinist propaganda.

'A Human Triumph that humbles Mankind'

A journalist, writing in one of the top newspapers in the UK, surpassed himself in the euphoria he experienced over the cloning of Dolly, "..In the sheepish gaze of Dolly from Edinburgh, awesome possibilities glitter. We can imagine, just a little, how it must have felt to be a Tuscan Jesuit reading Galileo's Dialogue on astronomy, or a pious Londoner settling down 250 years later with a first edition of *Origin of Species*."[12] The reason for his euphoria was that he really believes geneticists have begun to reveal how much is determined in our genes, and that in gaining control of our genes, human beings are gaining control of their own destiny. E. O. Wilson, the founder of the discipline of sociobiology, that purports to explain all human behaviour in terms of the natural selection of genetically determined behavioural traits, was quoted in the same article, describing the human brain (presumably human consciousness) as "an exposed negative waiting to be slipped into developer fluid". And, "The print is the individual's genetic history, over thousands of years of evolution and there is not much anybody can do about it."

In the same vein, Jonathan van Bierkom, professor of genetics at the University of Colorado commented, "After all, if you believe in the selfish DNA theory - the evolutionary imperative to propagate one's gene - then this is the ultimate."[13] Richard Dawkins - arch neo-Darwinian and genetic determinist, famous for the utterly banal idea that human beings are nothing but automatons acting under the influence of their 'selfish genes' whose only imperative is to replicate - also declared himself delighted. He confessed he would like to be cloned himself. He would love to watch a tiny copy of himself grow up. And fundamentally, he asked, was that not what we were all after when we had children? "So instead of watching a mixture of yourself and your partner's genes playing on the swings, you could watch the unadulterated you."[14] It was all too predictable that a letter in support of cloning, circulated on the Internet, was signed mostly by philosophers, together with those three arch-genetic determinists already mentioned: Richard Dawkins, E.O.Wilson and Francis Crick.

"Now we can reproduce ourselves without sex...", Andrew Marr continued triumphantly in his article,[15] chiding both "religious fundamentalists" and "open-eyed liberals" for calling attention to eugenics (while admitting that they had a point), but citing with approval novelist Fay Weldon's tongue-in-cheek comment that nature hasn't done such a good job that we can't improve on it, and that it is rather primitive of us to be so fearful of ourselves. It would definitely be a sin, he said, to use political authority to ban new thinking or new research. Tom Wilkie, a science journalist and now senior policy analyst with Wellcome Trust, was quoted as saying that moral attitudes evolve and that, up until 1950, it was illegal and considered immoral to use the corneas of dead people for transplant.

If people like Wilkie and Marr cannot tell the difference between using human corneas for transplant and cloning a human being, then we have not only descended into complete moral relativism but have also substituted Science for God. There is an underlying attitude that Science is, indeed, beyond reproach, that it can never be wrong, while "moral attitudes" or ethics are infinitely negotiable and evolvable. So, let us examine the science to see if it bears out the claims that have been made for it. To begin, let us look at how the cloning was done, and the claims that were made by scientists in a scientific journal, as opposed to those that appeared in the popular media.

What do the Experiments actually say?

In the 'cloning procedure', cells from an adult sheep's udder are cultured until they reach a 'stationary state' and cease to grow or divide. A cell is taken from the culture and fused with an egg from another sheep from which the nucleus has been removed. This allows the nucleus of the cell, containing the genetic complement of the first adult sheep, to substitute for the egg's genetic complement. The egg then starts to develop *in vitro* and, after making sure that it is developing normally, is transferred into the womb of a surrogate mother sheep who carries it to term. Out of a total of 277 embryos created in this way, only 29 developed sufficiently 'normally' to be transplanted into foster mothers. And of those 29, only one 'Dolly' resulted.[16] In the same series of experiments, cells were also taken from an early embryo and a foetus with which cloning was attempted. Of the 172 embryos created from foetal cells, 3 live lambs were born, one of which was very weak at birth and died soon afterwards. A total of 385 embryos were created from the cells taken from the embryo, and these gave rise to four live lambs, two of which were delivered by caesarean section. Thus, the overall success rate was no larger than 1%.

Actually, neither the idea nor the technique is new. Extensive experiments of this kind were done in the frog in the 1960s by John Gurdon's group in Oxford, and the axolotl by other developmental biologists. *In no case, however, did the scientists involved claim they were creating clones. Far from it, for they knew they were doing no such thing.*

The intellectual motivation for the experiments came from a deep problem in developmental biology. Organisms, no matter how complex, typically start development from a single fertilised egg cell which goes through successive cell divisions to produce many cells. These cells then undergo a hierarchical process of determination to form different organs and, later on, to become progressively differentiated into distinctive nerve cells, skin cells, liver cells and so on. There are two related questions which nuclear transplantation experiments enabled developmental biologists to answer. First, when cells become determined to form different organs, does the process involve irreversible changes in the genetic material carried in the nucleus, so that it loses the ability to form other organs and other cells? This is the problem in the 'totipotency' of the fertilised egg and its single copy of the genetic material which has the potential to become all the different organs and cells of the adult body. The second question is whether cell differentiation involves irreversible changes in the genetic complement carried in the nucleus of the cell.

It is significant that a scientific paper published in *Nature* did not claim that Dolly had been cloned. It was entitled, "Viable offspring derived from foetal and adult mammalian cells". Cloning was claimed, however, in the press releases and official comments passed on to the public. The implication of their claim was that the viable offspring, Dolly, contained the original "genetic blueprint" intact, so that adult cells could be used to produce another organism like the original.

In the earlier amphibian experiments, many developmental abnormalities resulted, and the furthest any embryo developed was to the juvenile, tadpole stage. However, by repeating the nuclear transplant serially - that is, taking cells from the first nuclear transplant embryo and transplanting the nucleus into a second egg cell - it was found that adult frogs could be created, most of which were infertile and abnormal in some way.[17] In one set of results cited, a total of 3546 nuclear transplants were done, using cells grown from adult frog skin.[18] The success rate of the first transplants to produce tadpoles was 0.1% - in other words, the failure rate was 99.9%. Serial transfers improved the success rate to 12%, but these tadpoles came from those 0.1% that had developed to tadpoles on the first transplant and were therefore pre-selected. And even the 'successes' showed varying degrees of abnormality.

The technique of nuclear transplantation was actually invented by King and Briggs in the 1950s.[19] They, and others subsequently, carried out extensive series of experiments. Some of the main conclusions arising from these experiments are as follows:

1. The developmental capacity of transplanted nuclei to support development decreases with the increasing age of the donor cells.
2. The reduced development capacity of the nuclei is irreversible, so that it is propagated over serial transplants and may involve DNA changes, such as chromosomal damage, as well as other alterations.
3. The developmental abnormalities resulting from the nuclear transplants experiments show no correlation to the kind of cells used.

Thus, there is no evidence that the original 'genetic blueprint' remained intact in any cell, except those obtained in the very earliest stages of development when the number of cells in the embryo could be visibly counted. Nevertheless, in summarising their results, Gurdon stated, "The main conclusion to be drawn from the experiments summarised in this chapter is that the nuclei of different kinds of cells in an individual appear to be *genetically identical.*" (my italics)

Gurdon's claim was not supported by the data and contradicted the subsidiary conclusions made just before. This was surely someone who was trying to salvage the accepted dogma that genes (DNA) do not change in development, only the expression of genes, *in the face of evidence to the contrary.* This misreading or misinterpretation of evidence is now familiar in the long history of genetic determinism. You will come across yet more instances in the next chapter.

The only real novelty involved in the Roslin Institute experiment was that it was done in sheep, and they succeeded in obtaining an apparently healthy live-birth without serial nuclear transplants. The interpretation of the results in the *Nature* paper was more cautious. Although it did not comment on the large proportion of failures, it stated, "The fact that a lamb was derived from an adult cell confirms that differentiation of *that* cell did not involve the irreversible modification of genetic material required for development to term" (my italics)[20]. Later on in the same paper, it was suggested that a judicious selection of cells going through certain stages of the cell cycle might enhance the success rate, although further experiments would be required to define the optimum cell-cycle stage for nuclear transplant.

But why were these experiments attempted? They were attempted in the hope of replicating animals with proven performance within 'elite selection herds'. In addition, the technique enables rapid creation of transgenic animals simply by making transgenic cells in culture for nuclear transplantation. As an afterthought, the authors mentioned that the technique could also be used for the study of possible persistence and impact of developmental changes on the DNA during differentiation.

The Science is Seriously Flawed

The science is fundamentally flawed in assuming that an individual is determined entirely by its genetic make-up. Furthermore, this is not supported by the results of the nuclear transplantation experiments, for reasons I have already described. Many commentators in newspaper articles have pointed out that the clone is not identical to the original individual, on account of the different life experiences the clone will have. Even identical twins, which are more 'clones' in the strict sense of the word, are different individuals. However, there are other more specific scientific errors involved.

First of all, as described above, one cannot clone any organism simply from a cell taken from the adult organism. It cannot be done without the egg from the second sheep, which plays the key role in 'rejuvenating' and 'reprogramming' the nucleus introduced with the cell, erasing all the 'imprinting marks' and other modifications in its

DNA that make it a mammary gland cell and, most probably, changing the introduced DNA in other ways so that it is appropriate to be the genome of a fertilised egg at the start of development. In addition, the egg cytoplasm provides important cues for making the proper body plan characteristic of the species - something which is yet very imperfectly understood, despite the isolation of large numbers of genes affecting the determination of body plan in the fruit fly in recent years. The egg also provides the food-store, as well as the sub-cellular 'power-houses' or *mitochondria* that generate the energy intermediate, ATP, which is used in all the energy transformations necessary for growth and development. The mitochondria, as it happens, have their own complement of DNA, and each mitochondrion with its DNA is replicated independently in the cytoplasm, so that when the cell divides, each daughter cell will have the right number of mitochondria. Lineages of organisms can be traced through the mitochondrial DNA, and mutations in mitochondrial genes are involved in a number of diseases. No cell can live without mitochondria.

The really interesting aspect of this experiment is the role played by the egg cytoplasm, which is almost uniformly ignored by all commentators, reflecting the patriarchal bias in current mainstream science. Nuclear-cytoplasmic interactions are well-known; development cannot proceed if the nucleus and the cytoplasm are incompatible with each other.[21] Many characters are so strongly influenced by the cytoplasm that 'cytoplasmic inheritance' used to be a subject of its own before it was eclipsed by the general obsession with DNA since the 1950s.

Another scientific error is in assuming that the genetic make-up of all the cells in the adult organism is the same, and identical to the fertilised egg from which the adult has developed. This myth has really been exploded since the early 1980s by the discovery of the fluid genome. As I pointed out above, it was already refuted by the nuclear transplantation experiments in amphibians. Thus, it is a case of bad science to ignore, if not wilfully misread, the evidence. In fact, somatic cells (cells of the body apart from germ cells) accumulate point mutations and other changes - insertions, deletions, rearrangements, duplications, amplifications and so on - during the lifetime of the organism. Some of these mutations are implicated in cancer (see Chapter 13). These DNA changes may account for the low success rate of the cloning technique. All in all, the experiment is misguided. It is not the best way to generate identical clones, but to generate monstrous failures of development. It is irresponsible and unethical to claim otherwise.

The Animal 'Pharm'

Dolly focuses our mind on a disturbing trend in genetic engineering as applied to domestic livestock and laboratory animals. The transgenic technology as a whole is very inefficient and the rate of success very low. Many embryos have to be manipulated and discarded due to abnormal, arrested developments before a few animals are obtained carrying the transgene, as was the case in making Dolly. Even the few that carry the transgene may turn out very sick, though they were not meant to be. This is simply because gene insertion is random, and the introduced gene is bound to interact with other genes in the genetic background, as I pointed out in earlier chapters. The most publicised failure of this kind concerned pigs engineered with a human growth hormone gene to make them grow faster. Unfortunately, they were arthritic, ulcerous, partially blind and impotent.[22] The widespread use of mice, made transgenic with mutant human genes to serve as models of human diseases, is both ethically and scientifically questionable. Is it justifiable to create animals that are meant to be sick? The most notorious is the 'oncomouse' which was engineered to develop cancer, and which has been patented and licensed to be marketed by Du Pont, without success.[23] The major scientific flaw in the whole endeavour is, as I have already said, that single human genes are put into a totally different genetic background, and so it is extremely doubtful that they can act as genuine models of the human conditions. For example, mice engineered to carry a mutation in a gene that predisposes humans to tumours in the retina of the eye (retinoblastomas) did not show any such symptoms. Similarly, mice engineered to have Lesch-Nyhan disease turned out to be totally asymptomatic, while those manipulated to have Gaucher's disease died within a day of birth.[24] You will see in the next chapter that human genes can even give completely different effects in different human populations.

It is in the world of commerce that genetic engineers have dreamt up ever more exotic ways to exploit animals to dubious ends, and which carry the most hazards.

Much harm for so little benefit

One project which has gained considerable momentum is xeno-transplantation, the transplant of organs of other species into human beings. There is already a thriving market in the sale of human body parts, as graphically documented in Andrew Kimbrell's *The Human Body Shop.*[25] Where once organ donors offered the gift of life without any recompense to those in need, body parts are now commercial commodities in the open market. It is estimated that the market for human organs may be worth $6 billion per year in the U.S. alone.[26] The biotech company, Imutran, based in Cambridge, is one of the world leaders in producing pigs with human genes to overcome the immune reactions that lead to rejection of transplanted organs.

Another major development is the engineering of domestic livestock such as sheep and cows to secrete useful drugs in their milk. The same laboratory and biotech company partnership that produced Dolly had already, in 1990, produced 'Tracy', the transgenic sheep that secreted huge quantities of the human protein, alpha-antitrypsin, in its milk.[27] This protein is secreted mainly from the liver in human beings, and is an inhibitor of elastase which breaks down a connective tissue protein. The deficiency of alpha-antitrypsin is associated with emphysema - obstruction of airways of the lungs. Tracy was made with a *large* segment of human genomic DNA surrounding the human gene, as previous experiments showed that the gene by itself was not expressed at a high enough level to be commercially viable. The researchers did not know what was in the rest of the DNA, but inferred that it must contain a regulatory sequence for high expression. However, Tracy's transgene was not stable, and failed to be passed on to her offspring. This was presumably the main reason why 'cloning' was attempted, in order to preserve the characteristic of Tracy without going through the process of germcell formation, which is where the transgene(s) can most easily get lost. Transgenic instability is a major problem in livestock as it is in crop-plants (see Chapter 9).

There are at least two aspects to consider in the 'pharming' of animals. The first is the ethical concern regarding animal welfare. Is the suffering of the animal justified by the amount of 'good' involved? The second is the question of whether it is safe. According to Richard Nicholson, editor of the Bulletin of Medical Ethics, the whole transplantation programme in rich countries adds 0.003% to life expectancy, which amounts to about a day. The use of xenotransplantation, assuming it works perfectly, will only increase life expectancy to 0.02%.[28] Clearly the good this programme can bring is extremely limited. The same may be said for pharming drugs in milk. The animal is made to lactate early with hormone treatment, and thereafter kept lactating permanently in order to keep up production. The protein the animal has to produce is *in addition* to all the normal proteins in her milk which, in the case of Tracy, is more than twice as much protein as in ordinary sheep milk. So she is under permanent metabolic stress. The protein has not yet been approved for clinical trial, so any good it might do is still unknown.

The hazards involved, however, far outweigh any potential benefits. A moratorium has now been imposed by the regulatory body in the UK on clinical trials of xenotransplants because of the recognised possibility that pig viruses can cross into human beings (see Chapter 13). As I have already pointed out in a number of chapters in this book, the mere act of transferring genes between unrelated species with chimaeric vectors is sufficient to facilitate the generation of new pathogenic viruses by recombination. Eukaryotic genomes are full of proviruses and related elements (see Chapters 8 and 13) which can help exogenous 'crippled' vectors mobilise and recombine with them. Recombination between exogenous and endogenous viruses are strongly implicated in cancer pathogenesis in many mammalian species (Chapter 13). The large segment of human DNA transferred to transgenic sheep like Tracy and other species is particularly worrying, as researchers have no idea at all what other sequences are in the extra DNA. It is very likely to contain human proviral sequences that could recombine with sheep sequences to generate new viral pathogens. Yet, the current trend towards constructing transgenic animals is to transfer large segments of the human genome to the other species. Scarcely a month after the announcement of the creation of the first artificial human chromosome in cultured human cells,[29] Japanese scientists made the first transgenic mouse line containing an entire human chromosome.

After all that, is the product safe to use? In the case of alpha-antitrypsin, which is marketed as a purified product, it may be possible to make sure that other protein contaminants, especially the prion proteins responsible for scrapie in sheep, BSE in cattle, and

Creutzfeldt-Jakob disease in humans, can be excluded, as well as viral particles. The animal, it is claimed, will be kept in a special disease-free environment. It is admitted, however, that "...practically speaking, it is impossible to ensure that production animals retain the same disease status from day to day. Certain viruses are endemic to particular livestock species in all parts of the world, and sub-clinical infections could go unnoticed."[30] The same company, PPL, is hoping to produce novel cow's milk for human consumption, claiming to use 'BSE-free' cows from the United States, and hopes to get approval for doing so. After all, it says, human blood is not currently checked for prior contamination. It expects "less-stringent guidelines in instances in which the products are for oral use."[31] As I have described in detail in Chapter 9, viruses and even naked DNA may survive passage through the gut and, from there, get into bacteria in the gut, and also *into many kinds of cells in our body.*

Conclusion

The cloning and 'pharming' of livestock, the creation of transgenic animals for xenotransplantation and to serve as animal models of human diseases are all scientifically flawed and morally unjustifiable. They also carry inherent hazards in facilitating cross-species exchange and recombination of viral pathogens. These projects ought not to be allowed to continue without a full public review.

Notes

1. Andrew Marr, *The Independent*, Wednesday, February 26, 1997, p. 17.

2. *The Guardian*, Wednesday, February 26, 1997, p.6.

3. Andrew Marr, *The Independent*, Wednesday, February 26, 1997, p.17

4. "Scientists scorn sci-fi fears over sheep clone", *The Guardian*, Monday February 24, 1997, p. 7. Also in a report and interview on the Eight O' Clock News, BBC Radio 4, Feb. 24, 1997.

5. Reported in *The Guardian*, Wednesday, February 26, 1997, p.6.

6. "Roslin patents come under the spotlight", *Nature*, 15 May, 1997, p.217.

7. Quoted in *The Guardian*, Thursday, February 27, 1996, p.5 of supplement.

8. Quoted in *The Guardian*, Wednesday, February 26, 1997, p.6.

9. Reported in *The Guardian*, Saturday, March 1, 1997, p.7.

10. Quoted in *The Guardian*, Wednesday, February 26, 1997, p.6.

11. See Spallone, 1992, Chapter 8, for an excellent critique of reproductive biotechnologies from a feminist's perspective.

12. Andrew Marr, *The Independent*, Wednesday, February 26, p.17, 1997.

13. Quoted in "Fearful symmetry", *The Guardian*, Saturday, March 1, 1997, p.1.

14. Quoted in "Fearful symmetry", *The Guardian*, Saturday, March 1,1997, p.1. Andrew Marr, *The Independent*, Wednesday, February 26, 1997, p.17.

16 Wilmut *et al*, 1997.

17. See Gurdon, 1974.

18. Gurdon, 1974, p.24.

19. King and Briggs, 1955.

20. Wilmut *et al*, 1997, p. 810.

21. Moore, 1955.

22. "And the cow jumped over the moon", *GenEthics News*, issue 3, pp. 6-7, 1994.

23. "The onco-mouse that didn't roar", Charles Arthus, *New Scientist*, 26 June, 1993, p.4.

24. See Lee *et al*, 1992 and Jacks *et al*, 1992 for "retinoblastoma gene" effects, and Davies, 1992 for "Lesch-Nyhan disease gene" and "Gaucher's disease gene" effects in transgenic mice. See also Wirz, 1997 for a recent critique of gene-centred biology.

25. Kimbrell, 1993.

26. "Phoney life on animal pharm", Denny Penman, *The Guardian* (Society) March 5, 1997, p.4.

27. Wright, *et al*, 1991.

28. Cited in "Phoney life on animal pharm" (Note 26).

29. "Human artificial chromosome constructed", *Nature Biotechnology* 15 May, 1997, p. 400.

30. Colman, 1996, p.641S.

Chapter 12

The Brave New World of Human Genetic Determinism

Ethical committees, by not questioning the scientific basis of the practices being considered, end up serving vested interests rather than humanity at large. The science of genetic determinism is both master and handmaiden to the industrialisation of eugenics. If screening is eventually going to be applied to 'predisposing' genes and to genes whose connection to dubious conditions is increasingly tenuous, we shall slip insensibly and quietly into an era of human genetic engineering dictated purely by corporate interests. This will lead to the exploitation of the sick and the gullible for profit and, at the same time, will give rein to the worst excesses of human prejudices.

Inherent in genetic determinist thinking is the tendency to be blind to the enormous variation that exists in all natural populations. Genetic tests are poor predictors for the condition of any individual because the genetic backgrounds of all the other genes are different. Genes associated with certain conditions in one population turn out to have no associations at all in another. Genes cannot be considered except in the context of the whole organism in its socio-ecological environment.

Genetics and Ethics

In November, 1996, a meeting was held in Nuremberg to commemorate the 50th anniversary of the trials of doctors who experimented on concentration camp prisoners during the Nazi era. During the meeting, the German section of the International Physicians for the Prevention of Nuclear War (IPPNW) called for a full debate in the German Parliament on the Council of Europe's proposed Bioethics Convention, claiming that it was not sufficiently restrictive to provide adequate protection for the mentally handicapped.[1] The meeting endorsed a declaration that if the draft Bioethics Convention were to be approved, there was a serious risk that Germany's dark history of human experimentation would repeat itself. This declaration has already been signed by more than 14,000 individuals, including representatives from the medical, academic and political communities, as well as groups such as charities for the mentally handicapped.

The declaration was initially drawn up for a meeting of the IPPNW's working group on the history of euthanasia, held in June 1995 in Grafenecker, a small town in southwest Germany where 10,000 mentally handicapped and psychologically ill people were gassed by the Nazis in 1940.

The psychologist, Michael Wunder, who drew up the declaration, argued that the draft Convention compromised the first article of the Nuremberg Code of 1947, which says that voluntary consent of the human subject in medical experiments is essential, and also specifies that anyone taking part in an experiment must be capable of giving consent. In contrast, the proposed Bioethics Convention would allow research to be carried out to determine the general mechanisms of a disease suffered by a person incapable of giving consent. This would be of no direct benefit for the patient, even though it might benefit others afflicted with the same disease. The declaration also argued that the Convention was too liberal towards human embryo experimentation and genetic screening. It took no stand on the issue of embryo research (this omission being especially glaring in view of the possibility of human cloning raised by subsequent events), but confirmed that genetic testing would be permitted for health purposes or for scientific research linked to health purposes, accompanied by appropriate genetic counselling. At the same meeting, the IPPNW also called for a moratorium on the development and use of new genetic screening tests until laws had been introduced protecting individuals against their possible misuse.

But Ludger Honnefelder, director of the Bonn Institute for Science and Ethics, and a member of the Council of Europe's expert committee responsible for drawing up the Convention, rejected those criticisms, claiming that the Convention had been wilfully misinterpreted with a political intention to "hinder genetics research". The Convention was approved by the Council of Europe's parliamentary assembly in September 1997, when the German representatives were the only source of significant objection.

Once again, the underlying assumption is that *science must go on*, with ethics being negotiated around it. The laws of nature, the reasoning goes, are ineluctable, for they are God-given; while the laws of 'man' must accommodate themselves to the progress of science and not hinder it.

China has already legislated for eugenics in a new law which came into effect in June, 1995.[2] It requires couples planning to marry to undergo screening for 'serious' hereditary diseases, as well as contagious diseases such as AIDS, sexually-transmitted diseases, leprosy and mental diseases including schizophrenia, manic depression and other major psychoses. If either partner suffers from infectious or mental illnesses, the marriage must be postponed. In the case of genetic diseases, marriage is only allowed if the couple agree to long- term contraception or sterilisation.

Genetic screening, including prenatal genetic screening and gene replacement therapy, are already widely available in the west. The British Government's Advisory Committee on Genetic Testing (ACGT) has produced a code of practice for companies selling genetic tests by post, but the code is not legally binding.[3] It restricts testing for carriers (heterozygotes) of recessive diseases, such as cystic fibrosis, and requests companies to notify it in case they want to sell tests for other types of genetic disorders. It says genetic data should be kept confidential, although samples may be transferred to third parties with the consent of the person whose sample it is. This permissive attitude is at odds with the recommendation of the Government's own Science and Technology Committee that commercial screening should be strictly regulated, on the grounds that "there is a very real danger that unscrupulous companies may prey on the public's fear of disease and genetic disorders and offer inappropriate tests, without adequate counselling and even without laboratory facilities necessary to ensure the tests are conducted accurately"[4].

Genetic discrimination and eugenics are privatised and depersonalised. They are also much more insidious than the state-sanctioned forms because they cannot be effectively opposed, as they are promoted under the banner of scientific progress and free choice. It is significant that Genetic Science and Industry ranks before Human Rights in the Science and Technology Committee Report, emphasising the importance of gaining access to the healthcare market in the USA and elsewhere.[5] Has no-one questioned the ethics of profiting from ill-health? In the same way, by not questioning the scientific basis of the practices being considered, ethical committees end up serving vested interests rather than humanity at large.

The science of genetic determinism is both master and handmaiden to the industrialisation of eugenics. It proliferates, like a virus, into every aspect of our lives. There is, admittedly, a better scientific rationale to hunt for traditional 'single gene' diseases - such as cystic fibrosis and sickle-cell anaemia - that are clearly shown to be associated with mutations in a corresponding gene. But screening for these diseases is not without problems, as I shall show later. The hunt goes on, however, for genes said to 'predispose' people to diseases such as cancer, diabetes, asthma, allergies, to conditions such as obesity, manic depression, schizophrenia, alcoholism, homosexuality, criminality, and even to attributes such as longevity, novelty-seeking and so on.[6] They have made sensational newspaper headlines but, behind the headlines, many of the genes have come and gone like will-of-the-wisps, because they never existed in the first place. Of the rest, all that remains to be said is that they give varying degrees of association with the given condition, which differ for different human populations.

"The effects of genes are complex, subtle and depend on interactions with the environment. In many cases, possession of the gene variants associated with a disease will only increase the risk and will not necessarily indicate that their possessor will suffer from the complaint. Even in the minority of cases where the possession of a defective gene inevitably leads to the development of a condition, it cannot predict its severity or, in late-onset diseases, show when it might appear."[7]

Nevertheless, tangible demand has been created for screening, not just for the single gene diseases, but for any other condition potentially on offer. A recent survey carried out in the U.S. found that 43% of those with genetic disorders were experiencing discrimination in employment, health and life insurances.[8] If genes make people susceptible or allergic to environmental pollutants, or to pollutants in their workplace, then employers will demand those forms of genetic screening for employment.[9] And why stop there? Geneticists claim to have identified a mutation in a gene encoding monamine oxidase inhibitor that causes aggressive behaviour, and may be implicated in Attention Deficit Hyperactivity Disorder (ADHD) in young children, Conduct Disorder in adolescents and Anti-social Personality Disorder in adults.[10] One scientist even suggested that six-year-olds diagnosed with ADHD might be saved from a criminal career if they were given prophylactic drug treatment.

The fact that some people may be able to tolerate poisonous chemicals does not make it ethical to make them work in places where such pollution is the norm. And branding a child a potential criminal on account of its genes is simply to relinquish responsibility for its care and proper upbringing. By a seemingly harmless, subtle shift in emphasis from the environment to the genes, we are saying effectively that it is all right for the environment to deteriorate and to be polluted, and that it is OK for communities to undergo social disintegration that drives adults to drink and crime, and children to misbehave. What's wrong, we are told, lies in people's genes. Therefore, what we need to do is genetically engineer people *not* to be susceptible, commit crimes, get ill and so on.

Such research is not only a drain on public resources, diverting them away from the real causes of society's ills; it is pernicious, on account of the ideology of genetic determinism that motivates it, and which it in turn reinforces. Let us examine the shaky foundations of the claims of genetic determinism and the difficulties it generates, beginning in the area of classical Mendelian genetic disorders, then going on to the more elusive human conditions.

Hunting for the Snark

Gene hunts are based on the simplistic assumption that there is a gene or a limited number of genes that can be pinpointed for every known human condition, however complex, and whether or not it is known to be significantly influenced by environmental factors. Recombinant DNA techniques now provide the means to follow the *association* of any given condition with a large number of genetic *markers*, for which there is variation, or *polymorphism*, in the population. These genetic markers are banding patterns produced on electrophoresis by pieces of DNA with unknown function, resulting from cutting the DNA in the human genome with specific restriction enzymes, or from amplifying specific sequences using the PCR chain reactions (see Chapter 3).[11] The hope is that one of those markers will be closely linked together on the same chromosome with the putative gene, *also of unknown function*, determining the condition, and so will be more likely to be inherited together with the gene. By screening members of those families in which the condition is known to occur, the marker most likely to be present together with the condition can then be identified. This technique has been successfully used to locate the gene in cases of so-called single gene diseases which, to put it in proper perspective, comprise less than 2% of all human diseases. (This, by the way, puts the upper limit to the validity of Mendelian genetics as an explanation of heredity.) In other cases, most of the markers turn out to have little or no relevance to the condition, when more data are analysed or other families are examined.[12]

The human genome is huge, containing some 3 billion base pairs, only a small fraction of which code for functional proteins and RNAs, or are involved in providing signals for making functional proteins and RNAs. Some estimates put the proportion of useful DNA at 1% to 5%, others, up to 33%. The rest, of unknown function, is referred to as 'junk DNA'. This is typical of eukaryotic genomes, with 'junk DNA' vastly predominating. No one knows why so much apparently useless DNA is carried around. The number of polymorphic genetic markers is enormous. By 1995, 10,468 of these had been mapped to locations scattered throughout the 23 human chromosomes. So, the idea of picking up functional genes by looking at their association with these markers is like hunting for a needle in a haystack, even assuming that the needle exists. Hunts for genes involved in single gene diseases have a much better chance of success because most of the effects *can* be attributed to a single gene, even if no-one has any idea of what that gene does. This is the power of 'reverse genetics' that recombinant DNA techniques enable geneticists to do. One first isolates the gene, then finds out what the gene does. Classically, it had to done the other way round - it was necessary first to identify the biochemical idiosyncrasy before locating the gene involved. Finding a gene is simple if the biochemical basis is known, or if classical linkage analyses and other studies have already supplied important clues as to where the gene might be located.

Cystic fibrosis was the first case in which a gene of largely unknown function was located by such methods. Its approximate location was already known from classical linkage studies, but it was still not an easy task. It involved extreme, at times bitter, competition between rival laboratories, and a couple of false reports before the gene was finally found in 1989.[1] There was an immediate proposal for screening and prenatal diagnosis. And the case for this is strong, for it is a single gene disease affecting a high proportion of the population.

The practical problem of screening is that the gene is very, *very* big, some 230,000 base pairs in length, with 27 exons, coding for a large protein that has 1,480 amino acids. A deletion of 3 base pairs, removing a single amino acid (in position 508) from the protein, occurs in 68% of the patients. But more than 400 different mutations have since been found[14] - and others will turn up on further screening - involving changes in different positions within the same gene. Only some mutations result in cystic fibrosis, or cystic fibrosis-like syndromes. The same mutations, moreover, may be associated with different symptoms in different individuals. It would not be feasible to screen for all the variants. Screening for only the commonest variant will not reveal whether the other mutations may or may not also result in cystic fibrosis in combination with the common variant. This sort of problem is shared by many other single gene diseases.

Not only can many different mutations in the same gene lead to the same syndromes, but so can mutations in one of several genes, as is found in a conglomerate of cranio-facial syndromes which includes achondroplastic dwarfism. Conversely, mutations in a single (tyrosine kinase) gene have been attributed to four different syndromes.[15] As mentioned in Chapter 3, there is really no such thing as a single gene disease.

At present, the sort of diseases where screening can give a definite result are those found in people with a known family history of the disease, and where, as in Huntington's chorea (which affects one in 20,000 people in mid-life) the mutation simply involves variable repeats of a trinucleotide sequence (CAG) within a particular gene, and the condition appears to be associated with 40 or more repeats - there being a rough correlation between the age of onset and the number of repeats greater than 40.[16]

What about the genes that 'predispose' towards cancer and other conditions? 'Breast cancer genes' have been in the news since 1994, when mutations in two genes, BRCA1 and BRCA2, were discovered which were said to 'trigger' breast cancer.[17] Both were thought to be 'tumour suppressors', though their real functions have remained elusive to this day. Other putative 'cancer genes' continue to be identified, as, for example, *BRCA3*,[18] and *BCSG4*,[19] there being such a big potential for the screening market. In the case of *BRCA1*, women who inherit one copy of the mutated gene inherit the predisposition to breast cancer, as, when the unmutated copy is damaged in the cell, that cell will turn cancerous. The frequency of this gene in the North European population is 0.0033 or 1 in 300.[20]

It is estimated that one in ten women in western societies is likely to develop breast cancer by the age of 85 and 25% of those will die from it. However, only between 2% and 5% of all cases are known to be hereditary, and the first two genes discovered, between them, account for only 80% of those cases known to be hereditary. *But 'predisposing' mutations in those particular genes do not predispose women to develop breast cancer if they do not already have a family history of breast cancer.* So, a positive test for *informative* mutations in either one of those genes, if you have a family history of breast cancer, will tell you one of two things. Either you belong to the 20% where it does not matter that you have the gene, in which case your chance of getting breast cancer is still one in ten, as in the general population; or you have an 80% chance of developing the disease. These tests will give no information, whatsoever, on the vast majority (95%) of breast cancer cases.

A genetic test for *BRCA1* is now being marketed in the U.S. To date, 254 mutations have been identified, of which 132 are unique and only 36 are 'disease-associated'.[21] The most common mutations, 185delAG and 5328insC, account for 11.7% and 10.1% respectively of all mutations shown. In the course of screening, other mutations are bound to turn up in the gene whose effects will be utterly unknown. It is relevant to mention here that, in a study to ascertain whether breast cancer may be induced by mammographic screening,[22] as has been widely claimed, 99% of mammographically-induced breast cancers were found to have occurred in women who were carriers of a breast cancer gene. This identifies at least one important environmental factor in the aetiology of the disease, even in its familial form.

The medical benefit of knowing about a predisposition to breast cancer, as for other conditions for which there is no cure, is unclear. It is more likely simply to result in increased anxiety, with some women even resorting to drastic prophylactic surgical removal of the breasts.[23]

'Predisposition' conceals the fact that important environmental factors are left out of consideration. This is the case not just in cancers but in such conditions as late-onset diabetes. This is characterised by resistance to insulin treatment, for which genes have yet to be identified, although it is claimed that between 2 and 8 genes may be involved. But, even if they are eventually identified, "there will still remain the daunting task of interpreting the way these genes predispose to a disease that is strongly influenced by external factors such as poor diet and obesity."[24]

Progress in identifying genes that predispose towards cardiovascular disorders fares no better. In the United States, a quarter of men and women under the age of 65 and about half of those over 65 die from cardiovascular diseases. Some cases - identified as suffering from familial hypercholesterolaemia - have been traced to a mutation in a gene, but most other cases are due to a general susceptibility and lifestyle. Even in familial hypercholesterolaemia, where a mutation in the cholesterol-carrying protein leads to the disease, the onset and severity of the condition is affected by diet.

Thus, in most cases, it will be impossible to interpret the results of a genetic screening test for the patient, or for the insurance company or prospective employer. The only people who stand to benefit in all circumstances are those selling the screening tests. But then, that is what the healthcare market is all about.

Complexities Confound Prognosis

The other major problem is that prognoses can vary a lot, even for so-called single gene diseases. A positive result in a screening test will not give an accurate prognosis of how the individual will fare. And, in the case of 'pre-dispositions', the individual may not develop the condition at all.[25] Even if one were to ignore environmental effects, the expression of each gene is entangled with that of every other. In other words, a prognosis will depend on the genetic background, consisting of all the other genes. Enormous variation exists in every known gene. Hundreds of mutants have already been uncovered, while many of the variants are essentially neutral, in having little or no deleterious effects. Thus, no two people on earth will have the same combinations of genes, except for monozygotic twins at the beginning of their lives. The uniqueness of the individual applies not just to his or her life experiences, but especially to the genetic make-up.

The variable prognosis is particularly relevant in conditions such as cystic fibrosis and sickle cell anaemia, which can be alleviated by treatments and appropriate management of crises. And one should never forget to mention the classic case of phenylketonuria (PKU).

This is a condition which in some, but not all, cases, resulted in severe mental retardation and early death before it was effectively treated by excluding foods from the diet containing the amino acid phenylalanine, which PKU sufferers cannot metabolise. Nevertheless, positive test results for PKU have already created social pressures on parents to abort the foetus,[26] whether they wish to do so or not. So-called 'therapeutic' abortions will soon be seen as *the* logical follow-up to positive pre-natal diagnosis, and no-one will be seriously looking for cures or treatments any more, except in the form of gene therapy, which creates its own problems (see next chapter). The major impetus now is to isolate the gene for the purpose of screening and then eliminating the affected before birth. And if screening is eventually going to be applied to 'predisposing' genes, and genes whose connection to dubious conditions is increasingly tenuous, we shall slip insensibly and quietly into an era of human genetic engineering dictated purely by corporate interests exploiting the sick and gullible for profit, while simultaneously giving rein to the worst excesses of human prejudices.

Inherent in genetic determinist thinking is the tendency to be blind to the enormous variation that exists in all natural populations. It ignores the fact that a variant which appears to predispose to a certain disease in one population may not have the same effect in another, simply because *the genetic background is very different.* Thus, cystic fibrosis mutations, in the Yemenite population, are found to be associated with a different syndrome - the bilateral absence of vas deferens - which can also result from some other unknown cause(s).[27] Significantly, in the same population, *none of those patients diagnosed as bona fide cystic fibrosis sufferers actually possessed the cystic fibrosis mutations.*

Another case in point was captured in a headline in the *Boston Globe* which proclaimed, "Genes Tied to Cancer in Jewish Women". This referred to the finding that about 1% of American Jews have the 185delAG variant of the cancer gene, *BCRA1*. The assumption that followed was that there was a 'Jewish genetic flaw' that might predispose Jewish women to breast cancer. There is no basis for such a conclusion. As Hubbard and McGoodwin[28] point out, the variant "may merely be a common form in which this gene occurs in the Jewish population". It does *not* suggest that Jewish women have an increased predisposition to breast cancer. This example serves, once again, to stigmatise Jews. And already, a biotech company has begun offering screening tests to Jewish women for a fee. When a screening *was* carried out on 108 Jewish women suffering from breast cancer, who also had a familial history of breast cancer, no more than 24 (23%) of them were found to have the mutation.[29] This is considerably lower than the figure of at least 40% found to be associated with mutations in *BRCA1* among Northern Europeans, the other 40% being associated with mutations in *BRCA2*.

An even more dramatic difference in association has been found for cases of *male* breast cancer.[30] Analysis of 54 patients showed that *none* of them had any *BRCA1* mutations. Two patients were found to carry novel mutations in *BRCA2*, but only one of the two had a family history of cancer.

One can confidently predict that, as more 'predisposing' genes are studied in other human populations, they will be shown to have different degrees of association, or none at all, to the corresponding conditions. It will be a case of "Now you see it, now you don't". This is the consequence of the non-linear complexity and interconnectedness of gene function, where diversity is a hallmark. Classical geneticists have repeatedly wrestled with the question of why there should be so much diversity in natural populations, and the debate goes on. The most immediate explanation is that genes and genomes are inherently mutable and fluid. Mutations in both germ cells and somatic cells are much more frequent than previously thought (see next chapter). Genes work together as an interconnected network, in which a lot of variations will have no net effect on the wellbeing of the individual concerned, i.e. they are neutral. Potential metabolic blocks can also be bypassed because of redundancy of pathways within the system.[31] Another explanation is that, while specific combinations of alleles may work particularly well with one another, some may give harmful effects in homozygous states. Thus, the sickle cell anaemia allele is widely thought to give protection against malaria in the heterozygous state *in the Afro-Caribbean genetic background*, while the cystic fibrosis allele is thought to give protection against childhood diarrhoea,[32] *in the North European genetic background.*

Classical genetics has long estimated that every individual in a population carries at least the equivalent of 5 lethal alleles in heterozygous forms. This means either heterozygous alleles in 5 genes, each of which will be lethal in homozygous form, or heterozygous deleterious alleles in many, many genes, equivalent to a total of 5 'genetic deaths', if all of them were made homozygous.[33] This is referred to as 'the genetic load' of the population. Population geneticists have long ago demonstrated that eugenics does not make good scientific sense. It makes even less sense in the light of the new genetics, when we know that genes and genomes are fluid and dynamic, that mutations are much more frequent than previously thought and that they can occur *in response* to environmental conditions, as was extensively reviewed in Chapter 8. Let us go on to deal with the more tenuous connections between genes and patterns of human behaviour.

Neurogenetic Determinism

Neurogenetic determinism purports to explain brain function and human behaviour in terms of genes. It "claims to be able to answer the question of where, in a world full of individual pain and social disorder, we should look to explain and to change our condition"[34] - to our genes. According to this theory, people are homosexual because they have 'gay genes' and people are violent because they have 'criminal genes'. Of course, the language is usually clothed in due sophistry, but the message is unmistakable.

Neurogenetic determinism is a direct descendant of neo-Darwinism and operates in the same way. As in the latter, the first step in the process of neurogenetic determinism is to invent a character (see Chapter 6). 'Aggression' is a familiar character invented by Konrad Lorenz,[35] who did most to 'explain' all animal behaviour as being the result of the natural selection of genes determining their behaviour. Aggression is now firmly established in the literature of sociobiology - the discipline that 'explains' all social behaviour, including that of human beings, in terms of the natural selection of genes determining social behaviour. Thus, it comes as no surprise that 'aggression' is used to lump together all kinds of behaviour from temper tantrums in children to a man abusing his partner or child. The description offered in the study which claims to have identified association of aggression with a mutation in the gene encoding monamine oxidase inhibitor is an example of the intellectual sleight-of-pen involved. The 'behavioural phenotypes' of the eight males in a family said to have the mutated gene include, "aggressive outbursts, arson, attempted rape and exhibitionism"[36] - activities carried out by subjects living in different parts of the country at different times over three generations.

"Can such widely differing types of behaviour, described so baldly so as to isolate them from social context, appropriately be subsumed under the single heading of aggression?"[37] Yet the 'evidence' provided by this paper is part of the argument employed by the current Federal Violence Initiative in the USA to identify at least 100,000 inner-city children whose alleged biochemical and genetic defects will, it claims, make them prone to violent crimes in later life.

Similarly, claims and counter-claims for 'genes for schizophrenia' have been appearing since 1988, and the story is not yet over. This condition affects one in every 200 people in the west at some time during their lives. They experience disturbances of thought and feeling, a belief that their actions, as well as thoughts and feelings, are under external control. They may experience visions, seeing, hearing or even smelling things that others cannot, and may be convinced of something for which there is no obvious justification.[38] But, not only is the hereditary status of schizophrenia questionable, "there is no general agreement about its cause, its diagnosis, its symptoms, its cure, or even whether it actually exists as such."[39]

There are currently at least four claimants to 'genes for schizophrenia'. One of them is located near a 'genetic marker for schizophrenia' based on measuring anatomical features of the brain by magnetic resonance computer tomography as well as diagnosis of 'schizophrenia-related', 'schizotypal personality disorder' and 'schizophrenia' - all lumped together in a single family.[40] Mutation in another gene, the human dopamine D-3 receptor, is said to display, in homozygous state, a "two-fold higher risk of schizophrenia",[41] while third and fourth candidates are those genes[42] - not yet identified - involved in the incorporation and removal of certain fatty acids into the phospholipids of the cell membrane. It is clear that, in each case, the connection is tenuous and weak, and not helped by lumping together different syndromes. Also, even if biochemical or anatomical differences do exist, they do not tell us whether these differences are the cause or the consequences of their conditions.

Nevertheless, a dominant hereditarian view of schizophrenia exists. It can be traced back to studies on identical twins carried out by Franz Kallman in the late 1940s and early 1950s. These studies claimed that a high level of *concordance* exists between identical twins - that is, where one twin suffers from schizophrenia, the corresponding twin is much more likely to suffer from it too, and conversely, where one twin does not suffer from schizophrenia, the other twin will also be free of the condition. Unfortunately, when re-examined by Richard Marshall in 1984, those studies were found to be riddled with distortion and fiction.[43] For example, Kallman himself made both the diagnosis of schizophrenia and the ascertainment that the twins were identical. Kallman was a committed eugenicist who extolled the virtues of eugenics and biological psychiatry. He referred to people with schizophrenia who did not need hospitalisation, as "disease trait-carriers". When his methods and data were challenged, he failed to produce clarification. Kallman's work is an example of how results can be distorted to fit an ideology, then uncritically accepted and used by a scientific community attuned to the same ideology. That was neither the first, nor the last of such episodes found in science, and genetic determinism has had more than its fair share. An even more spectacular distortion occurred with respect to the genetics of IQ (see below).

It is significant that, two years after the initial claim that a gene for schizophrenia had been isolated, the British medical journal *Lancet* carried a report of a study[44] suggesting that the improved health of mothers was a factor in the substantial fall in the number of people admitted to hospital for schizophrenia since the 1950s. But genetics and nutrition by no means exhaust the factors that may be involved in the genesis of schizophrenia. On the contrary, it is the totality of a person's experience of life in which the genetic and nutritional, the "social, cultural, spiritual and cosmological"[45] are all inextricably entangled, that counts. There is no single, isolatable cause that applies universally to all cases, genetic or otherwise. To persist in thinking that this is the case, and to pursue a course of action as though this were true, is to perpetrate untold violence upon the human spirit. And that is much of what the brave new world is about.

The Multigenic IQ Fraud

When hereditarians fail to pinpoint the gene or genes determining a condition, they retreat to polygenic, or multigenic inheritance. IQ, or intelligence quotient, is claimed to be such a character. Robert Plomin, who worked for many years in the United States on the genetics of IQ, has taken up a prestigious post in the Institute of

Psychiatry at the Maudsley Hospital in London. His research group is proposing to identify the genes involved in intelligence by means of genetic marker associations. This has worked successfully for schizophrenia, criminality and homosexuality. So, they must reason, why not for IQ? Plomin and his colleagues have applied to the Medical Research Council (MRC) for £1.8 million to take DNA samples from 10 000 children.[46] This has aroused considerable public opposition.

The IQ test was originally invented by Alfred Binet, for the benign purpose of identifying children who were not profiting from instruction in the regular public schools of Paris, so as better to help them. IQ, as he identified it, was strictly a ratio of performance, or quotient, between mental and chronological age. He did not, for a moment, suggest that his test was a measure of some innate, or fixed characteristic that could not be changed. To those who made this claim, his answer was, "We must protest and react against this brutal pessimism".[47]

In both the U.S. and the UK, IQ tests came to be used in a way diametrically opposed to that originally intended. Claims were made that these tests measured an innate quantity reflecting mental ability or intelligence, which was fixed by genetic inheritance at birth. Followers of Francis Galton - the Galtonian eugenicists - took control of the mental testing movement in all English-speaking countries. According to them, IQ measures were indicative not only of genetic differences in intelligence among individuals, but also between social classes and human races.

In England, the major translator of Binet's test was Cyril Burt, whose father was Galton's physician, and it was Galton's strong recommendations that secured Burt's appointment as the first school psychologist in the English-speaking world. From an early age, Burt was motivated by eugenic ideas, which he subsequently sought to confirm in voluminous publications. In both the United States and the UK, IQ tests have shunted vast numbers of working-class and minority children into inferior and dead-end educational tracks. The testing movement was also clearly linked, in the United States, to the passage of compulsory sterilisation laws aimed at genetically inferior 'degenerates' and 'imbeciles', which resulted in hundreds of thousands Americans being sterilised between 1924 and 1974.

Burt continued his eugenics research into the inheritance of IQ until he died in 1971, knighted in Britain and be-medalled by the American Psychological Association, his work an inspiration to further generations of hereditarians such as Arthur Jensen and Hans Eysenck. But the cracks began to appear soon afterwards in the eugenicist edifice that Burt constructed. His most impressive data to measure the heritability of IQ was a study on 53 pairs of separated twins, the largest on record, in which he claimed to have found strikingly high correlation between twins in their IQ, but no correlation whatsoever in their environments. Furthermore, in order to fit a biometrical genetics model, he also claimed to have measured the correlations for a considerable number of types of relatives - the only investigator in history to have administered the same IQ test to the whole gamut of relatives! And the correlations all added up to his satisfaction - allowing him to conclude that variation in IQ was completely determined by the genes, the environment having no effect whatsoever.

Let me digress for a moment to repeat what heritability estimates involve. As pointed out in Chapter 6, the heritability of a character is the proportion of the total variation of the character, in a randomly mating population, which can be attributed to the variation in the genes.

$$\text{heritability} = \frac{\text{genetic variation}}{\text{genetic variation + environmental variation}}$$

Thus, heritability is strictly a population measure, and says nothing about the proportion to which any person's IQ is determined, or fixed, by the genes. Heritability can be estimated by correlations between different pairs of relatives - from monozygotic twins who have all their genes in common, to grandparent/child pairs who have one-quarter of their genes in common, and so on. In order for estimates of heritability to be valid, the population has to be random mating. That means mating purely due to chance, so that there is no mating between close relatives, and those people with similar genes do not mate together more often than their frequency within the population would predict. This ensures that there is a sufficiently representative range of genetic variation in any sample taken. Another requirement for valid estimates of heritability is that there should be a sufficient range of environmental variation existing in the sample of people being measured. These requirements are very seldom satisfied in human populations. There is a high degree of social stratification which severely limits the range of environments experienced by the pairs of relatives. In other words, they are most likely to be drawn from the same social class, and hence do not experience different environments.

From the equation given above, it can be seen that, if there is no environmental variation at all, or when environmental variation is underestimated, then all, or most, of the variation will be due to genetic variation, and one will get an artificially inflated heritability of 1 or close to 1. On the other hand, if there is no genetic variation within the sample, or when genetic variation is underestimated, then the heritability will be equal to zero or nearly so. There are, of course, more complicated genotype-environment interactions that are not taken into account.

Psychologist Leon Kamin dropped a bombshell when he first called into question the authenticity of Cyril Burt's work in a presidential address made to a meeting of the Eastern Psychological Association in the United States in 1972. The implausibility of Burt's claims should have been noted at once by any reasonably alert and conscientious scientific reader. To begin with, Burt never provided even the most elementary description of how, when, or where his 'data' had been collected. ...He never even identified the 'IQ test' he supposedly administered to untold thousands of pairs of relatives...even the sizes of his supposed samples of relatives were not reported."[48] And how did he manage to get the IQ scores of adults? (IQ tests are normally administered only to children.) Burt had written that he relied on personal interview; but in doubtful or borderline cases, an open or a camouflaged test was employed. "The spectacle of Professor Burt administering 'camouflaged' IQ tests while chatting with London grandparents is the stuff of farce, not of science."[49] Numerous inconsistencies were found between successive reports on the same 'data'. By 1976, the medical correspondent of the London *Sunday Times*, Oliver Gillie, was reporting that he could uncover absolutely no evidence that Burt's research associates, the Misses Conway and Howard, had ever existed.[50] In his article, he accused Burt of having perpetrated a systematic scientific fraud, a charge subsequently supported by two of Burt's former students, Alan and Ann Clarke, now themselves prominent psychometricians. Later on, Burt's biographer, Leslie Hearnshaw, also had to concur. A review of Hearnshaw's biography had this to say,

"Ignoring the question of fraud, the fact of the matter is that the crucial evidence that his data on IQ are scientifically unacceptable does not depend on any examination of Burt's diaries or correspondence. it is to be found in the data themselves. The evidence was there...It is a sorry comment on the wider scientific community that "numbers ....simply not worthy of our current scientific attention" .. should have entered nearly every psychological textbook."[51]

But as Rose, Kamin and Lewontin remark, it was not just a sorry comment on the wider scientific community. "The fraud perpetrated by Burt, and unwittingly propagated by the scientific community, served important social purposes."[52]

These "purposes" are none other than the validation of eugenicist policies that persecuted minority races, and all other politically dispossessed classes.

With Burt's heritability studies discredited, the few remaining twin studies all suffer to varying degrees from the limitations of small samples. For instance, some twins were not really separated, and the environments experienced by those twins who were separated may have been essentially too similar. The most comprehensive studies to date have measured the correlation in IQ scores between parents and biological children who share half their genes, and between parents and adopted children in the same families who share none of their genes. No significant difference could be found, indicating that the connection between genes and IQ is extremely tenuous and, even more so, the connection between genes and intelligence.

No one doubts that genes are involved in intelligence, just as they are involved in every other aspect of the living being. *However, that does not mean that specific genes determine particular traits. That* is the reductionist fallacy that refuses to see the living being as an interconnected, entangled whole. To hunt for genes in the belief that organisms are a combination of traits, each attributable to particular genes, is to engage in Lewis Carroll's hunt for the snark, where the snark turns out to be a boojum.

Any project to hunt for intelligence genes, or genes for other equally dubious multigenic traits, is simply based on bad science that has already been thoroughly discredited, shown to be rotten to the core. It has no place in our society, for it can only serve to reinforce the genetic determinist, eugenicist ideology that inspires it.

Notes

1. See "German physicians warn of genetics risks", Alison Abbott, *Nature* (News) 384, 5, 1997.

2. See "China legalises eugenics", *GenEthics News*, issue 4, p.1, 1995.

3. See "New guidelines for postal genetic testing", *GenEthics News*, issue 15, p. 3-5, 1996.

4. House of Commons Science and Technology Committee Third Report. *Human Genetics: The Science and Its Consequences*, volume 1, Report and Minutes of Proceedings, p. xciii, HMSO, London, July 1995.

5. In the Human Genetics Report mentioned in note 4 above, Genetic Science and Industry ranks 6th among the topics considered, while Human Rights ranks 7th.

6. See Spallone, 1992, Chapter 7.

7. Human Genetics Report (note 4) p. xxiii.

8. "Survey finds high levels of genetic discrimination", *GenEthics News*, issue 15, p. 5, 1996.

9. See the excellent critique of genetic determinism by Hubbard and Wald, 1993.

10. "Controversy over genes and crime conference", *GenEthics News*, issue 5, p.3, 1995.

11. For a detailed account of polymorphic gene markers and gene hunting, see Jones and Taylor, 1995.

12. See Strohman, 1994.

13. Karem, *et al*, 1989.

14. See Abbott, 1996; Alper, 1996.

15. Mulvihill, 1995; van Heyningen, 1994.

16. Jones and Taylor, 1995. p.126.

17. Brown and Kleiner, 1994.

18. Seitz *et al*, 1997

19. Ji *et al*, 1997.

20. Bowcock, 1993.

21. Lenoir *et al*, 1996.

22. Denoteer *et al*, 1996.

23, "An update on the 'Breast Cancer Gene' ", Hubbard, R. and McGoodwin, W. GeneWATCH 10, pl0, 1996.

24. Abbott, 1996 p. 390.

25. See Summers, 1996.

26. See Hubbard and Wald, 1993.

27. See Augarten *et al*, 1994.

28. See note 23.

29. Offit *et al*, 1996.

30. See Friedman *et al*, 1997.

31. Strohman, 1994.

32. See Human Genetics Report, 1995, p. xci.

33. See Ho, 1987c.

34. Rose, 1995, p. 380.

35. See Lorenz, 1977.

36. Brunner *et al*, 1993.

37. Rose, 1995, p. 380

38. See Rowe, 1987, cited in Spallone, 1992.

39. Spallone, 1992, p. 190.

40. Shihabuddin *et al*, 1996.

41. Lundstrom and Trupin, 1996.

42. Horrobin, *et al*, 1996.

43. Marshall, 1984.

44. See Oliver Gillie, *The Sunday Times* (London), October 24, 1976.

45. Fernando, 1990, p.24.

46. Campaigners vow to halt search for 'intelligence genes', *GenEthics News*, issue 10, p.1, 1996.

47. Binet, 1913, pp.41-42, cited in Rose *et al*, 1984, Chapter 6, which is an excellent account of the Cyril Burt affair.

48. Rose *et al*, 1984, p. 102.

49. Rose *et al*, 1984, p. 102.

50. Oliver Gillie, *The Sunday Times* (London), October 24, 1976, cited in Rose *et al*, 1984, p. 104.

51. Book review of Cyril Burt: Psychologist by J.S. Hearnshaw, British Journal of Psychology 71, 174-175, 1980.

52. Rose et al, 1984, p. 106.

Chapter 13

The Mutable Gene and The Human Condition

The inherent mutability of genes in the human genome is associated with many 'genetic' and 'non-genetic' diseases. This dashes any hope eugenicists might have for 'purifying' human populations from harmful or undesirable mutations. It also increases the uncertainty of genetic screening. Current attempts at gene replacement therapy are uniformly unsuccessful, and are already posing unacceptable hazards for patients. The design of more aggressive gene transfer vectors introduces further risks from genetic recombination of vectors with endogenous viruses to generate infectious, disease-causing viruses. Recombination between exogenous and endogenous viruses is strongly implicated in many cancers in animals. Similar hazards also arise in the proposed use of modified viral DNA as vaccines and in xenotransplantation of organs.

The 'Clockwork' Cell

"Somewhere in the pile of plans that describe how to make a man is a page with the heading: 'How to make eyes'....In the short-sight gene this page has a misprint somewhere on it ...Somewhere else there is a page on 'How to make hair'...."[1] So begins a popular book on genetic engineering, written by a molecular biologist. All the detailed plans for making an entire human being, we are told, are contained in the genes, which are like files in a filing cabinet. We are then told that the cell, with its "mass of complex machinery" is like a factory, organised from a central office (the nucleus) where the files are kept. But the DNA is not just passive files.

"DNA transmits its message to the cell, forcing it to listen", the book continues, "the nucleus [containing the DNA] is the ultimate source of all decisions and changes of direction in a cell's existence."[2] It is clear that the author's image of the cell is a metaphor for the predominant social relationship in our society, with the boss (predominantly male) ensconced in the central office, making decisions, and *forcing* workers to do his will, in a strictly one-way information flow. This picture fits so comfortably, so reassuringly, into the consciousness of those who are near the decision-making end of the social scale that it is very difficult to relinquish. And that is the picture largely held by genetic engineers, despite the fact that it is contrary to all the findings in molecular genetics that have accumulated over the past two decades.

There is no doubt that mainstream biologists are an anachronism. They have been left far behind, as physicists, chemists and mathematicians have, one by one, ceased to see the world in terms of static equilibria and linear, clockwork mechanisms. Biologists are stuck in the mechanistic era, refusing to see the reality of organisms as irreducible wholes within which genes (and genomes) are mutable and mobile as they respond to their cellular and physiological *milieu* which is ultimately connected to the external ecological and social environment. Lewis Wolpert, Fellow of the Royal Society in the UK, who also chairs the Society's Committee for the Public Understanding of Science, writes, "Science is the best way to understand the world. By understand, I mean gain insight into the way all nature works in a causal and *mechanistic* (my italics) sense." That effectively excludes all of quantum theory from his scientific world view.

Mainstream biologists are clearly unaware that the new key to living organisation - in place of linear, one-way genetic determination - is non-linear, multi-dimensional *intercommunication.* To assume otherwise, in the face of the irrefutable mass of existing evidence, as genetic engineers are doing, is the stuff of bad science. It is to subject the public to unacceptable risks. Science itself must be placed under the closest scrutiny by ethical committees, alongside issues of eugenics and genetic discrimination.

In this chapter, I shall review evidence for the high degree of mutability of genes in the human genome, which is associated with many 'genetic' and 'non-genetic' diseases. I shall then show how current attempts at gene replacement therapy are being frustrated by cellular and physiological reactions which already pose unacceptable hazards to patients. The design of more aggressive gene transfer vectors introduces yet further risks, due to the genetic recombination of vectors with endogenous viruses, of generating infectious, disease-causing viruses. Recombination between exogenous and endogenous viruses is strongly implicated in many cancers in animals. Similar hazards may also arise in the proposed use of modified viral DNA as vaccines and in xenotransplantation of organs.

DNA Callisthenics and Histrionics

The usual textbook picture of the DNA double-helix, such as the one presented in Chapter 7, gives the impression of a static molecule that, rather like a bad boss, issues memos to the 'workforce' to do the work, but does no work himself. In reality, DNA is now known to be flexible and highly mobile. It has to be, in order to work properly within the cell.[4] Stretches of DNA can adopt a variety of different conformations or shapes, depending on the base sequence, the base composition, the immediate environment surrounding the DNA, and proteins which convey messages *to* the DNA from the cell, as the cell responds and adjusts to its external environment.

The helix is usually right-handed, but a left-handed helix is also known. The many helical forms are in dynamic equilibrium with one another. DNA can be bent, kinked or unwound, or become super-coiled to form tertiary DNA structures. Even *triplexes* - consisting of three strands wound together - can be found. Many of the structures are formed as DNA interacts with enzymes that replicate DNA, that transcribe DNA or recombine with it by cutting and rejoining different stretches. Different DNA structures are also formed as it binds to the plethora of protein transcription factors (see Chapter 8). All these processes are very complicated. DNA replication, for example, requires at least eight different enzymes and proteins, and errors are invariably introduced, as the copying is not exact.

DNA in the genome is also subject to chemical modifications by stray ionising radiation (X-rays and gamma rays), UV light, and metabolic or chemical mutagens. Damage or mistakes in DNA replication are repaired by many different DNA repair enzymes. In vertebrates, though not in invertebrates, DNA is systematically modified by methylation - the addition of a methyl group, CH_3 - to the cytosine or the adenosine base; a reaction catalysed by DNA methylases. Methylation tends to silence genes, i.e. prevent them from being expressed, and is part of the cell's armoury of defence mechanisms against foreign DNA, such as viruses, that become inserted into the genome. In addition, DNA is subject to a host of fluid genome processes. These rearrange, delete, amplify, mutate, as part of normal development, or in the course of gene expression; or due to unknown, random causes. Similarly, DNA is carried in transposons and viruses jump in and out of genomes, causing disturbances to gene activities and changing genome organisation in the process. All these are described in some detail in Chapter 8.

So, most of the time, DNA is involved in rather mild exercises or callisthenics but, during big crises, it will also go into spectacular histrionics giving large changes in organisation. Molecular and reproductive biologist, Jeff Pollard, refers to DNA as a "metabolic molecule".[5] Jeff was one of the first to recognise the full implications of the new genetics back in 1984. Thus, a high mutation rate is expected, for mistakes can arise due to failures of repair, or due to fluid genome processes. DNA methylation itself is mutagenic, as methyl-cytosine is converted to adenine by deamination - removal of an -NH_2 group - and repair is less effective in this case. It is estimated that about one-third of all point mutations in human diseases result from converting a methylated cytosine paired with guanosine, an mCG pair, by deamination to a TG pair.[6] *Xeroderma pigmentosum* is a particular human condition caused by the failure of one kind of DNA repair mechanism that removes defective dimers - two adjacent bases on the same strand joined together - as the result of UV damage.[7] Patients with this disorder are hypersensitive to UV light and suffer from a high incidence of skin and eye lesions, including cancers, when exposed to sunlight. There are mutations in at least 8 genes that can result in this condition.

Germline Mutation Rates

The overall mutation rates in germline genes estimated from mutations in known single genes that give rise to diseases are between 7 to 9 per 1,000 live births. Of these, one-third are recessive mutations in autosomes (non-sex chromosomes), one-third are dominant mutations in autosomes, and one-third are sex-linked mutations on the X chromosome.[8] These single genes are spread over the entire genome, associated with a wide variety of conditions, such as adenosine deaminise deficiency, Duchenne muscular dystrophy, cystic fibrosis, sickle cell anaemia, haemophilia and Huntington's chorea. These mutation rates are *minimum* estimates, because they already exclude those lethal mutations that kill the foetus before birth, and also exclude those that do not lead to recognisable phenotypic effects.

A comparable figure for mutation rates comes from DNA fingerprinting, which was discovered by British geneticist Alec Jeffreys in 1985.[9] This is based on identifying variations in sections of DNA where certain short sequences are repeated a variable number of times in different individuals. These are known as variable number tandem repeats, or VNTRs. The technique of DNA fingerprinting involves cutting the DNA with restriction enzymes, running the fragments on electrophoresis and probing the bands with the short repeats of interest. By employing different mixtures of cutting enzymes and probes, a very complicated pattern can be generated which is specific for each individual. The pattern is used in identifying criminals and in assigning parentage. Children generally have a mixture of bands from both parents. However, up to 1% show new bands which are not present in either parent, representing new mutations that have arisen in the germ cells of their parents. These mutations involve changes in the number of tandem repeats of short sequences - 10-15 base pairs long - interspersed throughout the genome, mostly in non-coding regions or 'junk DNA'.

Variable numbers of very short tandem repeats also occur in coding regions of genes.[10] The first example of these to be uncovered was the fragile X syndrome, which may be accompanied by severe mental retardation. The condition is named in reference to a distortion of the X-chromosome in the patient, involving two small pieces that seem to be breaking away from the end of the long arms. Fragile X syndrome shows a strange pattern of inheritance in that the severity of the disease can increase or decrease over two or three generations.

A gene has been discovered to be associated with fragile X syndrome, coding for a protein that binds RNA, and is particularly abundant in the nervous system. It is a complicated gene, 38 kb in length with 17 exons (see Chapter 8 for details on the interrupted gene). A region within the first exon has repeats of the triplet, CCG. Unaffected people have 6 to 60 repeats, while those with fragile X syndrome have up to several thousand, and the more the number of copies present, the more severe the syndrome tends to be. The number of repeats tends to change in successive generations. Large numbers of repeats cause excessive methylation and the gene is silenced.

Huntington's chorea is a similar disorder. It is a dominant nervous degenerative disease with variable age of onset, and is due to the multiplication of copies of a CAG repeat within the associated gene. Those with the disease have 40 or more copies of the triplet; individuals with 34 copies or less show no symptoms. There is a negative correlation between the number of repeats greater than 40 and the age of onset. Again, the number of repeats can increase in successive generations, leading to earlier ages of onset. For unknown reasons, the number of repeats is more likely to increase if the gene is passed on by the father rather than by the mother. More than one-third of all cases of the disease in which the gene is passed on by the father show an increased in copy number, sometimes up to 40 copies. Mutations involving changes in tandem copy number repeats arise during DNA replication, and is thought to be due to faults in recombination between chromosome pairs, so that one chromosome ends up with more copies and the other with less.

In general, mutations arise with each round of DNA replication. As sperms take many more cell divisions to mature, and hence more rounds of DNA replication, there are more opportunities for mutations to occur in the male germ-cells. Most new mutations have been found in genes passed on from the father. The ratio of paternal to maternal germline mutation rate is about ten to one.

Investigations were carried out to estimate the rate of germline mutation by looking at patients with haemophilia B, associated with mutations in factor IX required for blood-clotting.[11] Independent mutations were found in 95% or more of all families with severe or moderate disease. Analyses revealed 'hot spots' for mutiple types of mutations. One of these was the dinucleotide sequence CG, which accounts for 25% of the independent point mutations observed in this gene. Another hot spot accounts for 8% of independent mutations, which are deletions or insertions of 20 bp or less. Larger deletions of 50 bp or more also occur, and are associated with inversions of the gene sequence - the original sequence read backwards. In addition, other deletions are found within the coding regions where alternating purines and pyrimidines, such as GT or AC, occur. It is estimated that at least 22 new germline mutations arise in each person. This is a minimum estimate, as many mutations will go undetected if they do not have phenotypic effects.

Of especial interest is the author's claim that germline mutations for factor IX are due to "endogenous processes" rather than external mutagens. This claim is based on comparing them to somatic mutations in a cancer-related gene, *p53*, which are caused by exposure to environmental mutagens such as cigarette smoke, aflatoxin and sunlight. But instead of pursuing the logical line of thought to the *physiological* conditions that can lead to germline mutations, he offers the genetic determinist 'explanation' that it is due to a Darwinian foresight to 'optimise' future evolution by generating a sufficient quantity of random mutations so that some of them may be advantageous. He further suggests that cancer is nature's way of eliminating those whose mutation rates are too fast.

Mutations may also be caused by the integration of viruses, retro-transcripts (see Chapter 8) or transposable elements into the genome. More than 500,000 separate integration events have been found in the human genome. Substantial sequence homology exists between exogenous retroviruses and retrotransposons found in the genome. Retroviruses, as you will recall, are RNA viruses that replicate themselves by a process of reverse transcription of their own RNA into complementary DNA, which is inserted into the host genome where it directs the synthesis of many virus RNA genomes as well as the protein coat that packages the RNA into infectious virus particles. Howard Temin suggested that retroviruses integrated themselves into the genome, gave up their independent existence, and evolved into transposable elements and other degenerate relicts. On the other hand, it was also possible that the viruses had evolved from cellular transposable elements.[12] In general, retroviruses and transposable elements are host-species specific.

A recent finding that gives cause for concern is that a transposable element, called *mariner*, originally found in the fruit fly, and since identified in many insects and other arthropods,[13] is now found in primates including humans, where it leads to a neurological wasting disease, Charcot-Marie-Tooth syndrome.[14] The gene containing the element, the *CMT* gene, has a large duplication of 1500 kb, caused by unequal recombination between homologous chromosomes. This is thought to be due to the DNA cutting enzyme encoded by the *mariner* cutting in the wrong place. Although the copy of the *mariner* in the *CMT* gene has been disabled, an active *mariner* element has been found in yet another gene which could have helped the disabled copy do its damage.

How has *mariner* crossed so many species barriers? It is thought to have crossed the species barrier while integrated into the genome of a virus or some other pathogen. Each time *mariner* enters a new species, it is thought to jump around wildly, disrupting many genes, until it loses its own genes for mobilisation. But at least one copy in the human genome remains active. Does that mean it is a recent acquisition? How recent? This same element has been experimented on by genetic engineers in an attempt to construct transgenic anti-malarial mosquitoes to control malaria, a practice which has already been criticised as thoroughly ineffective and irrational.[15] It is also highly hazardous. *Has mariner spread from transgenic mosquitoes to human beings? If such transgenic mosquitoes are widely released, it is bound to result in further spread of mariner elements to human subjects.* Geneticists are hoping to construct 'universal vectors' that can be used in all medically important insects.[16] This would have disastrous consequences on human health, of which the researchers are totally oblivious, being solely concerned with the way in which insects might be genetically engineered.

The high rate of germline mutation dashes any hopes eugenicists might have of purifying human populations from harmful or undesirable mutations. It also increases the uncertainty of screening tests and compromises their ability to predict the individual's risks of suffering from genetic diseases and other conditions.

Somatic Mutations and Cancer[17]

Somatic mutations are thought to be particularly associated with cancer which is predominantly a non-genetic disease, in the sense that it is not inherited. Cancer manifests as uncontrolled cell multiplication, usually starting from a single cell that has undergone malignant transformation, after which its descendant cells in the cancerous growth form a clone that share the same genetic change(s). However, many different genes can be involved in different cancers in different tissues, even in a single individual.[18] It is claimed that six to seven mutations may have to accumulate in the same cell before it becomes cancerous, as is consistent with the steep exponential increase in the incidence of cancer with age. The death rate from colon cancer in the United States rises from almost zero at age 20, to 100 per million at age 60 and to nearly 300 per million at age 85.[19] However, somatic mutation rates do not show a sharp increase with age. As measured by the loss of an enzyme activity in white blood cells, the frequency of mutation increases linearly from an average of 1 in 10^5 cells at 20 years of age to 2 in 10^5

cells at age 60. This increase in mutation rate is largely attributed to a decrease in the effectiveness of DNA repair.[20]

The genes related to cancer come into several categories. The two major ones are *oncogenes* which promote cell growth and *tumour-suppressor genes* which constrain cell growth.[21] More recently, three other groups have been identified: the *antiapoptosis* genes which prevent cell death, the *antimetastasis* genes which prevent the tumour from spreading, and the multi-drug resistance genes responsible for resistance to drugs administered in chemotherapy (see Chapter 8).[22]

Oncogenes are derived by mutation from normal cellular genes that promote cell growth, the *proto-oncogenes*. Mutations arise from many different mechanisms, including breaks and rearrangements in chromosomes. A major cause of mutation is retroviral integration into or near any of the genes. These retroviruses can also carry oncogenes which they have originally captured from the cell. In some cancer cells, the normal proto-oncogene is simply over-expressed due to a strong viral promoter inserted near the gene.

Tumour suppressor genes block cell growth, and tumour is thought to develop when both normal copies are mutationally inactivated. About 20 different tumour suppressor genes have been identified so far.

One cancer-related gene involved in cell death is the *p53*, encoding a protein that normally brings the cell cycle to a halt when DNA damage is detected, so that the cell dies instead of progressing on to the cancerous state. Mutations in *p53* may allow the cell to continue with its damaged DNA, resulting in multiple mutations that eventually lead to cancer.

It is clear that many different kinds of somatic mutational changes can be associated with cancer. The question is often raised as to whether the mutational changes may be the effect rather than the cause of cancer. Indeed, cancer cells accumulate mutations up to 1,000 times faster than normal cells, and the mutations could have arisen *after* the crucial transforming event.

There is no doubt that environmental factors such as X-rays, gamma-rays, cigarette smoke and other chemicals increase risks of cancer, though increase in mutations rates are not as clear-cut. Survivors of the atom bombs in Hiroshima and Nagasaki showed an increased frequency of cancers, including leukaemia, breast cancer, and cancer of the lungs and digestive system, compared with controls who had not been exposed to the bomb. Similarly, death rates from lung cancer are 5 times higher for asbestos workers and 11 times higher for smokers than the general population. For asbestos workers who are also smokers, the death rate is 53 times higher, resulting from the two risk factors multiplied together. Cancer is second only to cardiovascular disease as a major killer in industrialised countries. In addition to exposure to physical and chemical carcinogens, studies have implicated many other factors that increase risks of cancer, such as diets rich in saturated fats and meat, stressful lifestyles which influence hormonal status, weak electromagnetic fields near high-tension power lines and other electrical installations in the environment.[23] These factors are not known to be associated with an increase in somatic mutation rates, raising the possibility that somatic mutations may not be the primary cause of cancer. This will be explored in more detail in the final chapter.

Gene Dreams fade into Nightmares

There have been over 100 clinical trials in gene therapy since 1990, and "not much of what has been trialed (tried) does work"[24] "Hundreds of people have been treated with gene therapy, but no one has been cured. Is it time for researchers to return to the lab?"[25] Harold Varmus, the Director of the U.S. National Institute of Health, which has spent about $200 million a year on gene replacement therapy programmes, told a Congressional committee in May, 1995, "While there are several reports of convincing gene transfer and expression, there is still little or no evidence of therapeutic benefit in patients, even in animal models"[26]

Two major forms of gene therapy have been tried so far. The first, *ex vivo* gene therapy, relies on taking cells - such as bone marrow cells - from patients, transforming them in culture with the missing gene carried in a suitable gene transfer vector, and then returning the transformed cells to the patients. The other, *in vivo,* method, is to deliver the gene in the vector directly into the patients. Many of the problems come from the gene transfer vectors. For transforming cells in culture, retroviral vectors are used, as they infect cells that are multiplying and can insert themselves into the cell's genome. But because they insert at random, they can easily cause cancer, as mentioned above. For gene therapy *in vivo,* vectors made from adenovirus are used, which do not integrate themselves into the chromosome and are less likely to disrupt the genome. An alternative is to use liposomes to deliver the genes without using vectors. This gets the genes into the cytoplasm but not the nucleus, so they cannot be properly transcribed and expressed.

However, patients were found to develop immune reactions against the vectors after the first dose. In clinical trials on cystic fibrosis, patients' airways become inflamed after breathing in aerosols of the adenovirus. One patient almost died. Subsequent research in rats showed that injections of the adenovirus directly into the brain appeared to be harmless at first. But when the same animals were injected with the adenovirus in the foot two months later, they developed severe inflammation in the brain.[27] This was a surprise to the researchers, as conventional wisdom has it that the brain is protected from the immune system by the blood-brain barrier, which is not supposed to allow macromolecules to pass through. It is clear that gene replacement therapy already poses unacceptable risks to patients, and threatens the survival of the programmes themselves.

Desperate attempts are now underway to salvage the gene replacement therapy programmes, as private companies have also invested heavily in developing this potential healthcare market. Researchers are seeking other kinds of vectors, one candidate being the insect virus, baculovirus, which has been found to invade mammalian cells (see Chapter 9). The other is the AIDS virus, a retrovirus that can infect non-dividing cells. The vector constructed from the AIDS virus had all the genes that the virus needs to duplicate itself removed, so that it could only be multiplied with cell lines that harboured the endogenous helper virus. The possibility had been raised that the disabled AIDS virus could recombine into a virulent form and cause AIDS, but at least one of the authors of the research has been undeterred. It is clear that the pressure is on to construct more and more aggressive gene transfer vectors, which carry even more inherent hazards, as it is, by now, well-known that viruses can recombine, and that human genomes already harbour endogenous proviruses and related elements ready to help disabled viral vectors mobilise and to recombine with them. Furthermore, recombination between exogenous and endogenous viruses is strongly implicated in the development of many cancers in animals, while culturing viruses can itself generate virulent variants from initially benign forms. In short, *contemplated programmes of gene replacement therapies are very likely to cause disease and generate virulent viruses.*

Hazards from Vector Mobilisation and Recombination

Retroviruses and retrovirus-like DNAs were first discovered in eukaryotic genomes in 1980 by Howard Temin,[28] along with retrotransposons and other elements which bear a strong resemblance to retroviruses (see above). Because of their sequence homologies to exogenous viruses, it can be predicted that recombination will occur between the endogenous elements and exogenous viruses. There is now experimental evidence of such recombination, which is directly involved in pathogenesis.

Feline leukaemic viruses (FeLVs) of domestic cats are transmitted by infection. They are capable of inducing either acute antiproliferative disease or, after a prolonged incubation period, cancers in the lymphatic system. It has been found that up to three-quarters of the exogenous viral envelop glycoprotein gene may be replaced by sequences from an endogenous virus to produce biologically active recombinant viruses, and that these recombinant viruses are involved in generating malignant lymphomas in infected cats.[29] Because the envelope protein is replaced largely by the endogenously encoded one, these recombinant viruses are able to escape immunity developed by the host against the exogenous virus. Many other recombinant viruses are generated, some of them being implicated in pathologies of the nervous system and of blood cells.

The mouse mammary tumour virus is transmitted by infection through milk - as exogenous virus - to susceptible offspring, or through the germline - as endogenous provirus. Exogenously acquired and some endogenous mouse mammary tumour viruses are found to be expressed at high levels in lactating mammary glands, and the endogenous viral RNA is packaged together with the exogenous viral RNA in the same virus particles. This has given rise to a recombinant virus, containing part of the gene coding for the envelope protein from the endogenous virus, which has been found in the mammary tumours in infected mice.[30]

Similar findings have been made in viruses infecting birds. A highly infectious avian leukaemic virus was isolated which, on sequence analysis, was found to be identical to the extensively characterised Rous sarcoma virus, except in one gene, *gag*. This difference accounts for its super-infectivity.[31] The evidence also indicates that the new virus has been generated by recombination between the exogenous Rous sarcoma virus and endogenous viruses *in the cultured cells*. It indicates that culturing viruses is already a hazard by itself. The researchers are developing this supervirus as a gene transfer vector for birds, in total disregard of the potential dangers involved.

Recombination between an injected virus and endogenous viral genes has been demonstrated in AKR mice carrying a provirus.[32] These mice develop spontaneous T-cell (white blood cell involved in the immune response) lymphomas between 6 and 12 months of age. The immediate cause of lymphomas is not the inherited provirus, but requires the recombination between the provirus, AKv and two or more additional endogenous viruses. These recombinant viruses, called mink cell focus-inducing viruses (MCF), were named for their ability to induce abnormal growth of mink cells, indicating their expanded host range.

Recombination can also occur between different variants of an exogenous virus in the course of viral proliferation. Drug-resistant HIV-1 variants can be isolated from patients undergoing prolonged anti-viral chemotherapy, and can also arise in culture by passage through increasing concentrations of drugs. Viral recombination has now been shown to be responsible for generating multiple resistances to a single drug and also resistances to two different drugs, AZT and 3TC.[33]

These findings have enormous implications for the development of cancers, and also draw attention to the hazards inherent in gene transfer vectors. Because endogenous proviral sequences are ubiquitous in all genomes, there are bound to be recombination events between the introduced vector and endogenous viruses. The findings also draw attention to the risks involved in culturing viruses for vector-production, which can itself generate new viruses with increased host ranges. The dangers of generating pathogens by viral recombination are real. Over a period of ten years, six scientists working with the genetic engineering of cancer-related oncogenes at the Pasteur Institutes in France have contracted cancer.[34] The evidence remains circumstantial, as it would be ethically unacceptable to obtain experimental data intentionally. In the course of culturing a human retrovirus (HTLV-II) associated with hairy-cell leukaemia in humans,[35] two infectious viruses with deletions and rearrangements of their genomes, which have increased abilities to transform normal cells into cancerous states, were inadvertently generated.

Another related area which carries similar hazards from viral recombination is in the development for pest control of live recombinant viruses, or vaccines

made of parts of viral genomes or live recombinant viruses. Australian scientists have genetically engineered a virus that stops mice from getting pregnant, and hope to use similar viruses to control mouse plagues in the wild.[36] Critics point out that such contraceptive viruses could cause an ecological disaster and should never be released, as they could infect and sterilise non-target species.

A vaccinia-rabies recombinant virus was dropped in edible bait in Belgium a few years ago to vaccinate foxes against rabies This was, ostensibly, to eradicate rabies in domestic animals that could infect human beings. As foxes and other wildlife ingesting the bait could carry another poxvirus, such as the cowpox virus, which circulates among wildlife, recombination could occur between the two viruses. A survey carried out in 1996 revealed that 63% of bank voles and 7% of wood mice species *do* carry antibodies against cowpox, indicating the presence of the cowpox virus. Nevertheless, *in direct contradiction of the evidence*, the researchers concluded that, "The risk of virus recombination in wildlife can therefore be considered to be extremely low."[37] The following disturbing incident with a rabies vaccine ought to serve as a warning. An outbreak of rabies in the Serengeti Wildlife Park prompted a team of vets to give 34 animals a rabies vaccine. Within 10 months, four vaccinated dogs were dead and, since 1991, there have been no sightings of any of the dogs at all.[38] Did the vaccine regenerate live rabies viruses by recombination and kill the dogs?

Recombination between a viral vaccine against Aujeszky's disease and a co-infecting virus to regenerate infectious Aujeszky's disease virus has been demonstrated in cell cultures.[39] Vaccines are, in any case, notoriously ineffective against viruses that mutate and recombine to generate new variants. Vaccines can cause the very diseases for which they are supposed to offer protection in immunologically-deficient individuals, as was documented in a measles outbreak among a fully-immunised secondary school population in Corpus Christi, Texas in the mid 1980s.[40] Within the UK, there have been nation-wide campaigns since the 1950s and '60s to mass-immunise children against infectious diseases including whooping cough, polio, diphtheria, measles and tetanus. These immunisations have effectively reduced the incidence of infectious diseases, but they have also caused hundreds of cases of deaths and permanent brain damage, well into the 1990s.[41]

The use of recombinant viral vaccines increases the likelihood of recombination and also potentially broadens the host range of the new virus(es) that may arise, as the viral genomes used in vaccines are often already chimaeric in construction. Vaccines produced in transgenic plants[42] for use in humans and domestic livestock may be particularly hazardous in this regard, as they have increased potential for generating new pathogenic viruses with extremely broad host ranges. There are two general strategies for producing vaccines. The first is to insert pieces of animal viral coat protein genes into plant viruses, such as the cowpea mosaic virus. The chimaeric virus is then multiplied in susceptible plants, which also synthesise the chimaeric viral coat protein. The second strategy is to create transgenic plants, transformed with the appropriate chimaeric vector containing the animal viral coat protein or those parts of it that are responsible for the immune reaction, so that the plant synthesises the protein continuously. The plant material is then fed directly to animals and human subjects for immunisation. Both these practices are based on creating chimaeric viruses or endogenous virus-like elements that already have broadened host ranges. Their ability to generate yet more viruses that attack a wide range of species should not be underestimated. As I pointed out in Chapter 1, such wide-host range cross-species viral pathogens may have already evolved, some for the first time. There is no indication that research workers themselves have taken this possibility into account.

One area where the possibility of generating new infectious agents by recombination has been taken into account, at least in Britain, is xenotransplantation - the use of animal organs, especially pig organs, for human transplantation. Robin Weiss, a professor at the Institute of Cancer Research in London, has shown that a pig retrovirus is capable of infecting human cells in the laboratory. And, once the virus has gone through one life-cycle in human cells, it is then able to infect a wide range of other human cells.[43] This is one of the reasons why the UK Advisory Group on the Ethics of Xenotransplantation Committee has imposed a moratorium on clinical trials.[44] Proponents claim that transgenic pigs can be raised 'pathogen-free' for the purpose of transplantation.[45] But it will be impossible to get rid of the endogenous proviruses and related elements which are ubiquitous in eukaryotic genomes. It is those elements that will pose the greatest threats for recombination. The same problem, to a lesser though still significant extent, is inherent in the use of transgenic farm animals such as sheep and cows engineered with human genes to produce pharmaceutically exploitable proteins, or in the armies of transgenic mice engineered with defective human genes to serve as models of human diseases. The exploitation of animals in this fashion has raised serious concerns about animal welfare (see Chapter 11), particularly as any benefit that can be gained from such reductionist models of health and disease are questionable, at best.

Checklist of Hazards in Human Genetics and Medicine

The many and varied applications of genetic engineering biotechnology in human genetics and medicine have special implications for ethics and health, for humans as well as animals. In this checklist, I have collected together issues raised in Chapter 11, 12 and 13:

a. Ethical implications

1. Genetic discrimination from diagnostic tests.
2. Negative eugenic practices in 'therapeutic' abortions.
3. Positive and negative eugenic practices in *in vitro* fertilisation and diagnostic techniques.
4. Marginalisation of women in the commercial control of reproductive technologies.
5. Possibility of immoral use of human embryos in 'pharming' and in providing tissues and organs for transplantation.
6. Immoral use of humans and human embryos for experimentation.
7. Negative impacts on animal welfare in 'pharming' practices.

b. Hazards to human and animal health

1. Risk of cross-species epidemics due to facilitated recombination between animal and human viruses in xenotransplantation.
2. Risk of cross-species epidemics due to facilitated recombination between animal viruses and endogenous viruses in the human genome in 'pharm' animals - e.g. Tracy the transgenic sheep.
3. Risk of severe immune reactions from vectors in gene replace-

ment therapy.

4. Risk of cancer from facilitated recombination between gene replacement vectors and endogenous viruses.
5. Risk of superviruses arising from facilitated recombination between viruses and cells in culture.
6. Risk of cross-species superviruses arising from facilitated recombination between viral vaccines and endogenous viruses in plants, animals and humans.
7. Risk of harmful mutations from cross-species transfer of transposable elements.
8. Risk of new iatrogenic diseases from new generations of genetically engineered drugs and vaccines.

Notes

1. Bains, 1987, p.15.

2. Bains, 1987, pp. 26-27.

3. Lewis Wolpert, who heads the Royal Society's Committee on the Public Understanding of Science, states, "Science is the best way to understand the world. By understand, I mean gain insight into the way all nature works in a causal and mechanistic sense..." (Wolpert, 1996, p. 9)

4. See entries in Kendrew, 1995.

5. Jeff Pollard, cited in Rennie, 1993.

6. See entries in Kendrew, 1995.

7. See Cleaver and Kraemer, 1989, pp. 2949-2971.

8. See entries in Kendrew, 1995.

9. See Jeffreys *et al*, 1985. There is a great deal of controversy over the use of DNA fingerprinting in forensic evidence because of the technical difficulties involved, and also due to the size of the human populations from which individuals originate. Thus, additional bands can arise from sample decomposition just as lack of bands can result from incomplete enzyme action. Individuals from small populations share many more genes than those from large populations, and their DNA fingerprint may look very similar. See Thompson and Ford, 1990, for a detailed discussion of the problems involved.

10. Weatherall, 1993.

11. Sommer, 1995.

12. Temin, 1985.

13. See Warren and Crampton, 1994,

14. "Doctor, there's a fly in my genome", Philip Cohen, *New Scientist* 9 March, p.16, 1996.

15. Spielman, 1994.

16. See Warren and Crampton, 1994.

17. Jones and Taylor, 1996, pp. 197-213, for a description which covers all aspects of cancer.

18. Elnatan *et al*, 1996.

19. Cairns, 1978.

20. Jones *et al*, 1994.

21. Thompson *et al*, 1991, Chapter 16.

22. Leong *et al*, 1995.

23. This subject is dealt with in many papers in Ho *et al*, 1994.

24. "There's a whole lot of nothing going on", John Hodgson, *Bio/Technology* 13, p. 714, 1995.

25. Coghlan, 1995, p.14.

26. Coghlan, 1995, p.14.

27. "Gene shuttle virus could damage the brain", Andy Coghlan, *New Scientist*, 11 May, p.6, 1996.

28. See Temin, 1980.

29. Roy-Burman, 1996.

30. Golovkina *et al*, 1994.

31. Bieth and Darlix, 1993.

32. DiFronzo and Holland, 1993.

33. Gu *et al*, 1995.

34. "Cancer at Pasteur", *New Scientist*, 18 June, p. 29, 1987.

35. Chen *et al*, 1983.

36. "Alarm greets contraceptive virus", Ian Anderson, *New Scientist*, 26 April, p. 4, 1997.

37. Boulanger *et al*, 1996, p.247, see also McNally, 1995.

38. See Pain, 1997.

39. Dangler *et al*, 1994.

40. Gustafson *et al*, 1987. See Spallone, 1992, Chapter 6, for an excellent account on the questionable benefits of vaccination.

41. Harriman, 1988; also "Face the facts", BBC Radio 4 programme, June 19, 1997.

42. "High-tech herbal medicine: Plant-based vaccines", Charles J. Arntzen, *Nature Biotechnology* 15, 221-222, 1997; see also, Dalsgaard *et al*, 1997.

43. Reported in "Pork that could give us the chop", David King, *The Times Higher Education Supplement*, September 13, p. 17, 1996.

44. "Human xenotransplants banned in UK", Barbara Nasto, *Nature Biotechnology* 15, 214, 1997.

45. Greenstein and Sachs, 1997.

Chapter 14

The New Age of the Organism

It is clear that we need a deep and sustained change in direction in all spheres of life, before the dreams of solving all the problems facing the world today by genetic engineering biotechnology turn into nightmares. Contemporary scientific approaches that focus on the organism, on wholeness and complexity, are more consistent with the scientific findings in molecular genetics and in other disciplines. Organic stability arises naturally as the result of a balanced ecology. This has large implications for health and disease, especially in the development of cancer and the emergence of virulence and new pathogens. Sustainable systems can similarly be understood in term of a theory of the organism, whose carrying capacity is not rigidly set by the physical limits imposed by the second law of thermodynamics, but depends on its internal dynamic organisation. This holistic, yet rigorous, perspective in contemporary western science is consonant with traditional indigenous sciences all over the world. Reductionist science has had its day. Let us reject the bad science that has served to exploit, to oppress, to obfuscate, and to destroy the earth and its inhabitants. Let us opt for a joyful and sustainable future beyond genetic engineering.

Life beyond Genetic Determinism

Genetic determinism and capitalist economic theory stem from the same roots in 19th century imperial Britain, which they served remarkably well. The ideology of competition and exploitation validates and extols the social reality that gave it birth, and this social reality is in turn shaped and propagated by the ideology posing as science. Together, they have succeeded, only too well, in conquering the world.

The ever-intensifying exploitation of human beings and of nature by the major economies of the world has far outstripped the rates at which renewable resources can be replaced and substitutes for non-renewable resources developed. This has resulted in widespread environmental degradation, accumulation of toxic wastes products and, hence, a downward spiral of ever-diminishing returns in production, poverty and misery for the vast majority of the world's population.

"The main enemy of the open society, I believe, is no longer the communist but the capitalist threat."[1] This statement is remarkable, if only because its author is George Soros, a well-known capitalist who has himself benefited a great deal from the system. In the article, he also openly criticises neo-classical economic theory and social Darwinism, and refers to the same mutually reinforcing relationship between mindset and reality that I have developed at great length in this book. "There is a two-way connection - between thinking and events..."[2]

In this final chapter, I shall outline some of the elements of alternative, contemporary scientific approaches that focus on the organism, on wholeness and complexity, which are more consonant with the scientific findings in molecular genetics and in other disciplines. Richard Strohman, Professor of molecular and cell biology at the University of California, Berkeley, has written a number of important critiques of genetic engineering biotechnology for top biotechnology journals. In the latest one,[3] he describes the coming revolution in biology not too dissimilarly to the one I shall describe here. He points out how anomalies within the genetic determinist paradigm are being merely swept under the rug by "expert but conservative elements within the mainstream" in order to rescue the paradigm. But their explanations for the behaviour of complex systems become so contorted and convoluted, and invoke so many genetic agents with their "interactive and co-dependent states" that they serve only to obfuscate rather than contribute anything to further our understanding.

I start by considering the new view of organic stability and its implications for health and disease, especially for the development of cancer and for the emergence of virulence and new pathogens. I then introduce some new ideas on sustainability and biodiversity which have been developed in detail elsewhere. The object is to move science beyond reductionism to a holistic, yet rigorous, perspective which is consistent both with contemporary western science and with traditional indigenous sciences all over the world.

The Stability of Organisms and Species

How do organisms and species maintain their stability when genes and genomes are so mutable and fluid? The conventional, neo-Darwinian explanation is that natural selection is always at work, selecting out those that are unstable and hence 'unfit', so that only those that are sufficiently stable remain to propagate offspring like themselves.

This explanation fails to account for the *responsiveness* of genes and genomes to environmental and physiological changes, as described in Chapter 8 and elsewhere throughout this book. The stability of organisms and species is dependent on the entire gamut of dynamic feedback inter-relationships extending from the socio-ecological environment to the genes. Genes and genomes must also adjust, respond and, if necessary, change, in order to maintain the stability of the whole. As said earlier, the keynote to a living organisation is intercommunication and mutual responsiveness throughout the system.

The stability of organisms is diametrically opposite to the stability of mechanical systems. Mechanical stability - which applies also to so-called 'cybernetic' systems - is a closed static equilibrium, maintained by the action of controllers, buffers or buttresses which return the system to fixed, or set, points. A mechanical system works in a hierarchical manner, rather like most non-democratic institutions. Organic stability, on the other hand, is a state of dynamic balance which is attained in open systems far away from thermodynamic equilibrium. It has no controllers and no set points. It is radically democratic, as it works by intercommunication and mutual responsiveness of all the parts, so control is distributed throughout the system. It has a 'steady state' which consists of a complex configuration of dynamic *cycles* spanning a range of temporal (and spatial) domains. These cycles, which are most familiar to us as a whole range of 'biological rhythms', are effectively coupled together in relationships of reciprocity and cooperativeness, so that balance is achieved over the system as a whole.[4]

Thus, the dynamic stability of the organism depends on *all* parts of the system being able to adjust and respond appropriately in order to maintain the whole. When the system is well-balanced within its socio-ecological environment, the entire range of dynamic cycles ensures that all parts of the system engage in a kind of perpetual return, and so genes and genomes are also dynamically maintained in constancy. As you have seen, a healthy system has various enzymes that 'proof-read' replicated DNA to correct errors and repair chemical and physical damages (see Chapter 13). There is also a process of gene conversion that maintains the constancy of DNA sequence in those genes *that are in use* (see Chapter 8). On the other hand, when the system is stressed, mutational and other changes in genes and genomes will take place which may alleviate the stress, as in the origins of so-called 'adaptive mutations' (Chapter 8). But those changes in genes and genomes may also throw the system out of balance, resulting in its complete failure. Deleterious mutational changes, therefore, may be symptomatic of the breakdown of organic stability rather than its cause. This is becoming more and more evident in the development of cancer, as we shall see later.

Some direct evidence for the importance of the system as a whole in the maintenance of the stability of its parts is seen in the enormous variability of isolated cells in culture, as compared to their constancy and stability within the organism. This kind of 'somaclonal' variation is ubiquitous for plant cells in culture (see Chapter 9). Similar variation exists for cultured animal cells. Cell biologist Harry Rubin spent nearly 20 years documenting the endless variation of mammalian cells arising in successive passages in culture, despite their supposed genetic uniformity. He has proposed a concept of 'progressive state selection' to account for the physiological stability of cells *in vivo* in the living organism, as opposed to their variability in culture. "This concept assumes that physiological constraints can select among the ever-fluctuating physiological states in cells, and that repeated state selections result in heritability of those states. These considerations focus attention on the living cell and its neighbours which provide the immediate environment for selection, and ultimately on the whole organism."[5] Rubin's explanation, in terms of selection by the physiological environment, is not too dissimilar to my proposal that they are intercommunicating, so that, in the ideal, "each part is as much in control as it is sensitive and responsive"[6] Each cell within the body will contribute to the physiological *milieu* of every other cell.

This state of affairs is best summarised by medical physiologist Guyton, " ..the body is actually *a social order of about 75 trillion cells* organised into different functional structures, some of which are called *organs*. Each functional structure provides its share in the maintenance of homeostatic conditions ...which is often called the *internal environment.* As long as normal conditions are maintained in the internal environment, the cells of the body will continue to live and function properly. Thus, each cell benefits from homeostasis, and in turn each cell contributes its share towards the maintenance of homeostasis. This reciprocal interplay provides continuous automaticity of the body until one or more functional systems lose their ability to contribute their share of function. When this happens, all the cells of the body suffer. Extreme dysfunction leads to death, while moderate dysfunction leads to sickness."[7]

It will be instructive to examine how cancer can develop as a failure in the physiological ecology of cells, rather than as the result of random mutational events.

The Epigenetic Origin of Cancer

There is increasing evidence that cancer is not a disease generated primarily by the accumulation of random somatic mutations, as is usually understood, although progression to an irreversible cancerous state may involve deleterious somatic mutations. The alternative explanation is that cancers have an *epigenetic* origin. Epigenetic, in the present context, refers to physiological/developmental factors at least once removed from the 'genetic', which is anything that leads to changes in the base sequence of genes and genomes.

The genetic theory of cancer first came into prominence in 1981 when three separate papers reported that DNA derived from long-term culture of a human bladder cancer could transform a mouse cell line, NIH 3T3. The responsible segment of the human DNA was similar to a gene carried by the Harvey rat sarcoma virus, and was named the *ras* gene. That was the first of the many oncogenes which were subsequently discovered (see Chapter 13). It turns out that NIH 3T3 cells are uniquely sensitive to transformation by the *ras* family of genes. Moreover, this cell line readily undergoes 'spontaneous' transformation when subjected to moderate physiological stress. "None of these findings proves that the alterations in tumour cell DNA actually caused the tumours; they simply show that genetic change is much more common in many, though not all, tumours than in normal tissue."[8]

Strong evidence against the genetic origin of cancer came from experiments carried out in two laboratories, one of which used X-rays to induce malignant transformation, the other using the carcinogen, methyl-cholanthrene.[9] Both studies found that most, if not all, exposed cells were altered in some way, so that their progeny had a higher probability of transformation than untreated cells. In other words, the *entire* population of exposed cells showed an increased probability of transformation, and this increased probability was inherited in subsequent cell generations. So, if one divides up the exposed population into several sub-populations, each of them will show essentially the same frequency of transformation. Moreover, if these are further subdivided and propagated, the same frequency of transformed cells arises in all of the sub-populations. Such high frequencies of transformation are also characteristic of spontaneous transformations induced by metabolic stress, and are not due to correspondingly high frequencies of mutations. Still more suggestive is the observation that clones of cells transformed by X-rays or by metabolic stress revert to normal when placed under optimal growth conditions.[10] These results are reminiscent of the high rates of reversion in early stages of development of malignancies.

A team of Soviet scientists have been studying spontaneous transformations of cells cultured from a number of inbred mouse and rat lines for several years. The cells produced sarcomas in the mice and rats from which the lines were established. They found that, on culturing cells from the tumours, they were able to obtain colonies that were fully transformed, partly transformed or not transformed at all. On six successive re-clonings of transformed colonies, the cells persisted in giving rise to all three kinds of colonies. The non-transformed cells, which had lost their ability to give rise to tumours in the animals, arose at high frequencies. Commenting on these results, Rubin writes, "The question arises...whether the underlying process in all tumours is fundamentally epigenetic in character. I know of no experiment that rule out this possibility."[11] The strong implication is that mutations are not the primary cause of transformation. Instead they may arise after the crucial transformation of cellular state, which is the result of response to physiological stress. As said earlier, progression to the irreversible state may then be associated with somatic mutations.

Mutations may result from Physiological Stress

I believe we have to take seriously the possibility that mutations in human populations - both somatic and germline - may be the result of physiological imbalance and ecological stress, as has already been clearly shown for other organisms. Mammalian cells in culture, bacteria and yeast cells, as well as plants and insects all show increased mutation rates under metabolic stress or stress from cytotoxic drugs (see Chapter 8). This perspective has large implications on where we should focus intervention in health and disease. It is not in eliminating bad genes, or replacing bad genes with good in gene replacement therapy. Nor is it in the development of more and more exotic, specifically targeted drugs. There are already too many drugs on the market, causing a host of iatrogenic (prescription-drug induced) diseases that constitute a major problem in public health.[12]

So long as people's physiological states are compromised - by air pollutants, toxic wastes, pesticide residues and infectious bacteria in water and food, by malnutrition due to poverty, by substance abuse and by stress due to social disintegration - high rates of mutations will occur, which will lead to further deterioration in health. It is time we redirected our efforts and resources to address the real problems facing our societies instead of creating more and more exotic and hazardous ways to exploit the sick, the miserable and the gullible. Health and disease are inextricably matters of social, ecological and physiological balance. They have no real reductionist solutions. Many enlightened physicians and members of the public are already practising a variety of holistic medicines based on treating the whole person in his or her socio-cultural *milieu*, and in stimulating the self-healing, regenerative powers of the organism. The Bristol Cancer Help Centre in the UK was among the first to adopt holistic healthcare in the treating and preventing cancer, and has had a successful history dating back to the 1980s, despite recent efforts to discredit its work by the reductionist establishment.[13]

The Ecology of Quiescent Microbes

There is a strong parallel between the ecology of microbes in the environment and the physiology of cells within our body. The ecology of so-called quiescent microbes is a relatively new area of study, but it is already revealing important insights into the factors that make microbes multiply out of control and become virulent.

The most interesting form in which quiescent microbes occur is in biofilms - thin layers of extracellular matrix secreted by the microbes which house a single species or multiple species of microbes. Biofilms are found coating solid surfaces in the environment - such as rocks and stones, gravel in the bottom of rivers, lakes and ponds, and the surfaces of aquatic plants. They may also be found coating the surfaces in our gut and our circulatory system. Until very recently, biofilms were thought to be homogeneous layers in which non-growing bacteria are entrapped. When they were eventually examined by non-destructive microscopic techniques, researchers discovered, to their surprise, that biofilms are structured communities of single species or multiple species living together. The cells are disposed in aggregates around water channels through which convection currents flow, so that nutrients and metabolites can circulate.[14] In multi-species biofilms, mixed species micro-colonies are found where cells of metabolically cooperative species are juxtaposed so they can benefit from exchanging substrates and end products. The term 'quiescent' is not appropriate, as chemical probes indicate that most of the cells are metabolically active. The cells in a biofilm are effectively enjoying a kind of multi-cellular life served by a circulatory system. They are metabolically active but non-proliferating, just like most of the cells in our body.

Further studies have revealed that the same bacteria can live within biofilms or in free-living plankton. But the two states are associated with the expression of different genes as well as distinct morphologies. The physiology of biofilm bacteria is profoundly different from that of their planktonic counterparts. They live in large numbers of different microniches within multi-species communities forming stable colonies that persist indefinitely in the environment, where they are notably resistant to antibiotics. However, they also effectively prevent secondary colonisation by invading bacteria. This is particularly relevant for the normal complement of bacteria that live as biofilms in our gut. Studies have shown that biofilm species effectively prevent colonisation by invading pathogens.[15]

Observations on a wide variety of natural ecosystems have established that the vast majority of bacteria in most aquatic environments exist within biofilms in a non-proliferative, but metabolically active, phase, where exchange of genetic materials continue to occur. These biofilms probably play a large role in the recycling of nutrients for higher organisms, and contribute to maintaining the stability of the ecosystem as a whole. It appears very likely, therefore, that the so-called quiescent state is the ecologically balanced state, also for the microbial communities, and that they only enter the proliferative phase during times of stress. These may involve an influx of new pollutants or antimicrobials, of nutrients such as phosphates and nitrates, or some other major ecological disturbances. The emergence of new virulent pathogens is strongly associated with the destruction of ecosystems by encroaching industrialisation. Bacteria and viruses lose their natural hosts and become virulent in human populations whose physiological states are already compromised by malnutrition and poverty.[16] The ecological stress that gives rise to proliferating pathogens is analogous to the physiological stress that causes cancerous cells to proliferate out of control in the human body.

If that is the case, then the strategy of controlling infectious diseases during the past 50 years may be entirely misdirected. That was, indeed, the conclusion of the Harvard Working Group on New and Resurgent Diseases (see also Chapter 3). Reductionist science has failed yet another reality test.

It is the 'warfare with nature' mentality that has brought about the current public health crisis in infectious diseases and antibiotic resistance. There is an important moral in the way the microbes are winning the war over us. They are telling us that it is wrong to see the world in terms of isolated bits and pieces, to see linear chains of mechanical causes and effects. Nature is fluid and dynamic and thoroughly interconnected. The microbes do not compete, each against all the rest. On the contrary, they engage in rampant cooperation, sharing even their most valuable assets for survival against the onslaught of selfish, myopic human beings.

I cannot help feeling that warfare with human beings is not what microbes want to engage in either. Virulence and antibiotic resistance are signs of ecological stress, for them as for us. There is evidence, for instance, that bacteria which produce antibiotics also produce the enzyme for inactivating the antibiotic, and that what we call an antibiotic may actually be their sex hormone which signals the bacterium's readiness to mate (see Chapter 10). So, what we have done, with our profligate use of antibiotics, is to turn them into thoroughly promiscuous, sex-mad fiends. We have created, and will continue to create, an horrendous array of rogue strains of microbes in their midst. I imagine that they, too, would much rather things were different and that they could live peaceably with us in a balanced ecological relationship, as they once did long, long ago. It is high time we put this 'warfare with nature' mentality behind us and started learning in earnest how to live sustainably and healthily with nature.

A Theory of the Organism and the Sustainable System

The new genetics, in transcending the old reductionist paradigm, reaffirms the ecological wisdom of traditional indigenous peoples all over the world, who have practised sustainable agriculture on the understanding that the biological nature of each organism or species is inextricably linked to its environment, and depends ultimately on the entire ecosystem consisting of all other organisms. It is clear that intensive agriculture, in its new genetic engineering version, is just as unsustainable as the old, and carries new hazards inherent to the technology (see Chapter 9). The alternative is not to abandon science, as some disillusioned environmentalists in the North seem to be advocating, nor is it a wholesale return to 'traditional' methods. It is important to stress that so-called traditional methods have also evolved through the ages, as knowledge has accumulated, and it is simply Eurocentric arrogance to deny the existence of science in cultures other than the Northern European.

I made it clear at the outset that Western science is also *not* the monolithic reductionism that mainstream establishment scientists have been encouraged to project to the public. It is in the nature of all sciences to doubt, to explore and to evolve, in the light of new evidence, *so that one can gain a reliable knowledge of nature that will best enable us to live sustainably with her.* It is on that basis that I wish to outline a theory of the organism, within the framework of contemporary western science, and then to show how it may contribute to the understanding of sustainable systems.

Thermodynamic Limits to Growth

The notion of a sustainable system carries with it the idea that there are limits to growth and consumption. Goldsmith[17] and Rifkin and Howard[18] were among the first to draw attention to those limits, in particular, the physical limits imposed by the second law of thermodynamics, which says one can never get 100% efficiency in natural processes - in other words, there are no free lunches in the world. In order to live, human beings extract energy and materials and transform them. In the process, useful energy invariably degrades into a disordered, unusable form, *entropy*, while materials turn into waste products. So, there are physical limits to how fast these processes of extraction and transformation can take place before the system becomes overwhelmed by entropy and toxic wastes.

However, physical limits are simply ignored by mainstream neoclassical economic theories,[19] which all begin with non-physical parameters - preferences, technologies and distribution of incomes. These non-physical parameters are taken as given, while the physical variables of quantity of goods produced and resources required are adjusted to fit an equilibrium or, more often, a constant rate of growth. Sustainable development, by contrast, treats physical parameters as given, setting limits to the size of the economic system that can be supported. That is the basis of the ecological concept of *carrying capacity*, which reflects the size of the economic system and some optimum rates of transformation of energy and resources compatible with sustaining the human beings within the system.

But, how is the carrying capacity related to the rates of transformation of energy and resources? How is it related to the manner in which resources and energy are used, i.e. to the mode of production?[20] How is it related to the complexity of the system, say, in terms of diversity of production, the division of labour and so on? Which factors determine the carrying capacity of the system? Such questions are central to the issue of sustainability. And it seems to me that they are not sufficiently addressed by merely invoking the 'entropy law'.[21] I shall show how the physical limits are by no means static and given, but are dependent on the way the system is organised, thus providing a basis for effective human intervention and choice.

A Sustainable System is a kind of Organism

A starting point is to regard a sustainable system as a kind of organism. One can see that a healthy organism is the ideal sustainable system - an irreducible whole that develops, maintains and reproduces, or renews, itself by mobilising material and energy captured from the environment. The organism is sustainable precisely on account of its anti-entropic tendencies. It is paradigmatic in terms of the efficiency with which it uses energy and resources. In that respect, at least, it has as yet no equal in any machines that our most advanced technologies can produce. So it is in understanding the thermodynamics of the living system that one may learn how to overcome and minimise entropic dissipation and decay.[22]

The results of my studies over the past ten years suggest that the anti-entropic tendencies of the organism depends on its ability to *store* incoming energy, to make this energy circulate effectively within the system to do work before it is dissipated; that it is the *intricate* structure of the system which determines how long the energy can remain in it, how efficiently work is done, and hence how much energy is effectively stored within the system. It turns out that living systems have a dynamic structure that *maximises* energy storage and *minimises* energy dissipation. What does this dynamic structure consist of?

First of all, it consists of activities organised in *cycles*. Cycles involve perpetual return and account for the dynamic stability of the system. Secondly, the activity-cycles are nested one within another, like Russian dolls, spanning an enormous range of space-times from the very fast (nanoseconds) to the very slow (weeks or months), and from the very local to the increasingly global. In other words, the system has a deep space-time differentiation. Thirdly, its activities are all *coupled* together in a *symmetrical, reciprocal* way. This is the most subtle relationship to appreciate. It entails *both* local autonomy *and* global cohesion. Energy-yielding activities are directly linked to energy-requiring ones, in such a way that they can readily exchange places. In other words, there is a cooperative give and take: parts in deficit can draw on those in surplus, and the roles can be reversed, so energy is diverted to wherever it is most needed at all times. The net result is a dynamic balance, so the whole system is sustained.

This state of affairs is encapsulated in Figure 14.1. where the circle represents the entire life-cycle of the system, and the line the flow of energy that feeds the life-cycle. The life-cycle consists of the many subsidiary nested cycles of activity that enable it to achieve a net balance, especially of entropy production, represented by the equation, $\Sigma \Delta S = 0$, while the necessary minimum dissipation is exported to the outside, as represented by the equation, $\Sigma \Delta S > 0$. In this way, the organism effectively sets itself free from the *immediate* constraints of the second law, to become an autonomous system with a certain degree of independence from its environment. It is so organised that a full range of coupled cyclic, non-dissipative processes feeds off the dissipative flow. Consequently, the system maintains its organisation, and there is always energy available within the system for all its vital activities.

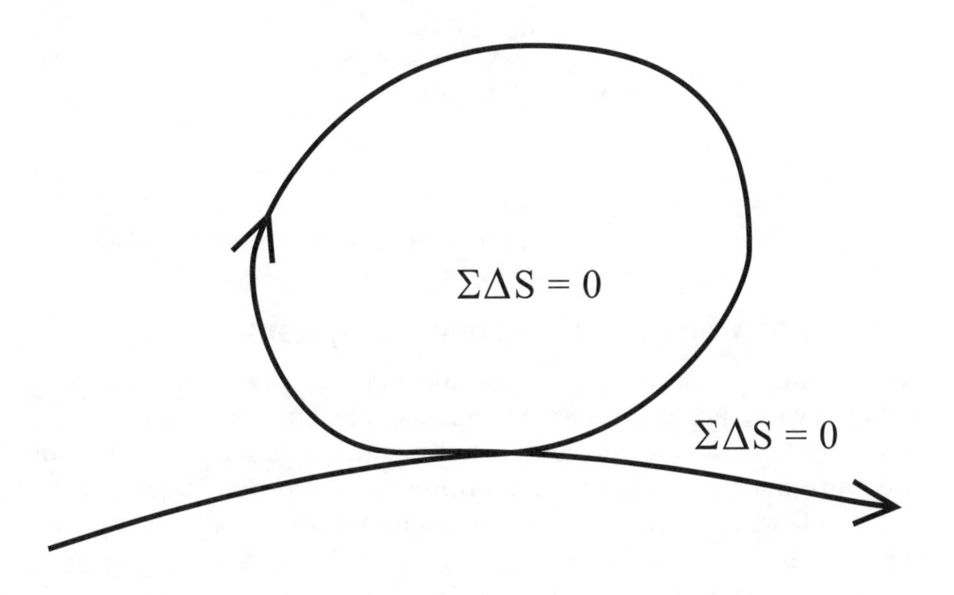

Figure 14.1. The organism can be understood in terms of cyclic non-dissipative processes feeding off irreversible dissipative processes.

The internal organisation, or intricate, dynamic structure of the living system, therefore, holds the key to its success. An intuitive representation of the intricate structure of the life-cycle is given in Figure 14.2. It consists of nested sub-cycles of different sizes which are dynamically coupled together. As you can imagine, the more sub-cycles there are and the better they are coupled, the longer the energy is effectively stored within the system and the less is dissipated as entropy.

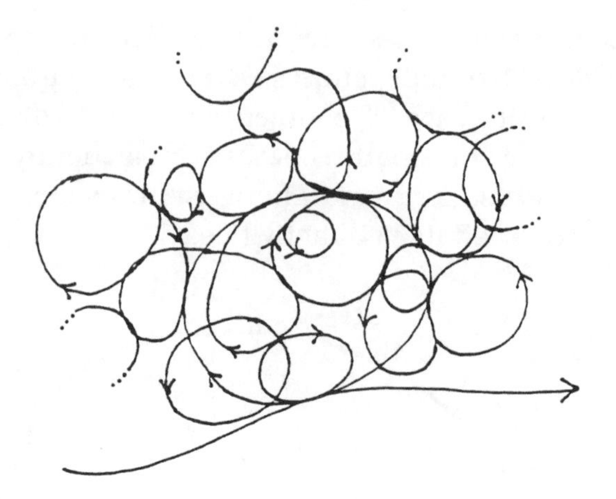

Figure 14.2 The intricate structure of the life-cycle consisting of many coupled sub-cycles.

A Model of a Sustainable Economic System

In modelling a sustainable economic system, one has first to remember what an economic system is, and what it is for. An economic system is a society organised around a mode of production. There are many possible modes of production in diverse cultures of the world, which differ among themselves with respect to distributions of resources and income, degrees of equity and emphasis on cooperation over competition.[23] Unfortunately, mainstream economic theories of the west are based exclusively on the capitalist mode of production with its sole emphasis on competitiveness, and that can make one lose sight of the purpose of an economic system. An economic system is, first and foremost, a society of people bonded by social contract to make their living together by using and transforming resources, and has as its purpose the achievement of a good life for all. It is, therefore, in everyone's interest to have a healthy economy.

The economy is, to first approximation, an open system through which resources extracted from the 'source' - the ecological environment - flow to a 'sink' - the most immediate mental picture of which is the municipal dump. Various commodities and services are exchanged or traded between 'source' and 'sink', and 'values' are added in processes of manufacture, creative acts of art or artisanship, whose equivalence to energy or otherwise need to be fully justified and explicated, along with such qualities of life as happiness, health, contentment and wellbeing, not to mention clean air, nutritious food, comfortable shelter from inclement weather and an unpolluted environment.

Like an organism, the economic system may be conceptualised in terms of cyclic, non-dissipative exchanges or transformations of energy and resources, coupled to the dissipative flows or wastage due to death, depreciation and other entropy-generating, irreversible processes. Because energy and resources come ultimately from the ecological environment, it makes sense to embed the economic system properly in its ecological setting (Fig. 14.3). This is fully in accord with Hermann Daly's proposal to "view the economy as a subsystem of the ecosystem, and to recognise that "while it is not exempt from material laws, nor is it fully reducible to explanations by them."[24]

Figure 14.3 makes it clear that the ecological environment is also concept-ualised as a self-sustaining organic system of cyclic, non-dissipative processes coupled to the dissipative, one-way energy and material flow. To what extent is this justified? Lovelock's Gaia hypothesis[24] proposes that the entire earth is a self-organising, self-regulating system maintained far from thermodynamic equilibrium under energy flow. In these respects, it is indeed like an organism. The most conspicuous sign of the earth's self-regulating property is the constancy of its atmosphere, which is a highly non-equilibrium mixture of gases. The atmospheres of Mars and Venus, by contrast, are equilibrium mixtures of spent, or exhaust, gases that reflect their lifelessness.

Let us say that the economic system depicted in Figure 14.3 is the global economy - which current World Trade Organisation negotiations are aimed at establishing. Then it will also have an intricate structure encompassing many local economic systems. Ideally, the intricate structure of the global economy should look like the many nested sub-cycles that make up the organism's life cycle (Fig. 14.2). Each local economy in turn, in, say, Britain or Malaysia, would have its own intricate structure of coupled cycles, self-similar to that of the global economy (again, like Figure 14.2). (This is the property of *fractals* - dynamics generating space-time patterns that fall between the 1, 2, or 3 usual dimensions we are used to - which characterise many living patterns and processes.) It is clear that local economies are coupled to the global through imports and exports of materials, human beings and capital. If the entire global system is to be sustainable, there has to be a proper balance between the local and the global, the same kind of reciprocal, symmetrically-coupled relationships that one finds in organisms.

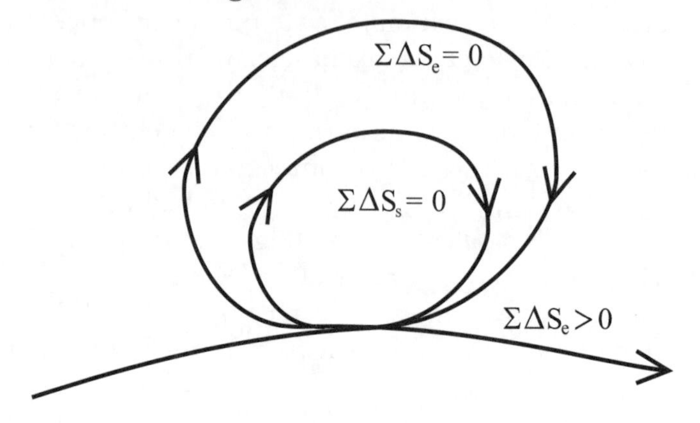

Figure 14.3 The coupled flows of the economic and ecological cycles in a sustainable economic system.

In the context of the planetary ecosystem, it is now generally recognized that human activity has had a far from benign effect. This has prompted the establishment of a 'Geophysiological Society' for the study of planetary health.[26] I suggest that the present model of the organism may begin to offer some diagnostic criteria of health.

A healthy organism is one that maximises the non-dissipative cyclic processes and minimises the dissipative. This gives a set of criteria for recognizing a healthy, sustainable system.

Diagnostic criteria for a sustainable system

- Maximise non-dissipative cyclic flows
- increase in energy storage capacity
  (carrying capacity, biomass)
- increase in efficiency of energy utilisation
  (effective coupling in energy transfer)
- increase in dynamic closure
  (cycles)
- increase in space-time differentiation
  (energy residence time, diversity)
- increase in balanced flows of resources and energy
  (symmetrical flows)
- increase in reciprocal coupling of processes
  (free energy conservation)

Minimise dissipation (minimum entropy production)

The diagnostic criteria for sustainable systems listed above are inter-linked, so that once one is identified, the others are very likely to exist as well. Some support for these criteria is that they are similar to those that Eric Schneider and James Kay,[27] have identified for mature, established ecosystems. These authors have made significant progress in understanding the thermodynamics of ecosystems, although their conceptual framework differs from the one presented here. They compared the data collected for carbon-energy flows in two aquatic marsh ecosystems adjacent to a large power-generating facility on the Crystal River in Florida. One of them, 'stressed', was exposed to hot water effluent from the nuclear power station which increased the temperature by 6°C, the other, 'control', not so exposed, was otherwise in similar environmental conditions. They found that the stressed system captured 20% less energy, made 20% less efficient use of the energy captured, had 50% fewer cycles and 34% less biomass than the controls. These findings suggest that the control system captures, stores and uses energy more effectively, has a greater carrying capacity in terms of biomass and a higher degree of dynamic closure (cycles), as consistent with the diagnostic criteria listed.

Schneider and Kay have also drawn attention to some interesting measurements made by Luvall and Holbo[28] with a NASA thermal infrared multi-spectral scanner, which assesses energy budgets from the air of terrestrial landscapes. Their data showed that the more developed the ecosystem, the *colder* its surface temperature. The interpretation by Schneider and Kay is that mature ecosystems are more effective in 'degrading energy'. An alternative interpretation, consistent with the thermodynamic model presented here, is that the mature ecosystems have a greater 'energy storage capacity', which is a well-defined thermodynamic concept. This greater energy storage capacity is related to its higher degree of space-time differentiation or intricate structure, greater dynamic closure (cycles) and more effective coupling of cyclic flows, achieving increased internal entropy balance so that the system exports less entropy or heat. In other words, the dissipative branch is minimised while the non-dissipative cycle is maximised in a mature ecosystem.

This model can throw light on many aspects of an economic system which are described in some detail elsewhere.[29] Here, I wish to concentrate on certain key features.

Money, Energy and Entropy

The first thing that comes to mind whenever the word 'economics' is mentioned is money. It is indeed money that makes the economists' world go round. So, it is all too easy to equate the circulation of money in real world economy with energy in the living system. It has even been fashionable, in biochemistry text-books, to regard the universal energy transducing intermediate, ATP (Adenosine Triphosphate), as 'energy currency'. However, money is by no means equivalent to energy. Perhaps a better analogue to energy in real world economy is affection, trust and goodwill, in return for which, originally, a gift was a token. Later on, people traded goods or services, which again depend on trust and goodwill, if the equivalence in energetic values is to be maintained. The big problem arose with the introduction of money, which is only arbitrarily related to the real value of things and services.

The result is that all money is not equal, and today's major economic difficulties arise from treating it as though it were. The flow of money can be associated with exchanges of real value or it can be associated with sheer wastage or dissipation. In the former case, it is more like energy flow, in the latter case, it is pure entropy.

Transactions in the financial or money market are purely for the purpose of extracting and concentrating 'wealth' away from real production. Money is thereby created, which is completely decoupled from the creation of value. As a result, as David Korten points out, "the financial or buying power of those who control the newly created money expands, compared with other members of society who are creating value, but whose real and relative compensation is declining"[30] It is estimated that for every $1 circulating in the productive world economies, $20 to $50 circulate in the economy of pure finance. This money represents pure uncompensated, positive entropy, or pure dissipation, in our scheme.

The other important source of uncompensated positive entropy is to be found in the unequal terms of trade consistently imposed by Northern powers on developing countries in the South. This has happened during both colonial and post-colonial times, with raw materials being extracted from the South with far too little compensation being paid, while, at the same time, manufactured goods are sold back to them at too high a price. Because the economic system depends ultimately on the flow of resources from the natural environment, which has its 'natural ecological economy', entropic costs can be incurred either in the economic system itself, or in the ecosystem. Thus, when the cost of valuable non-renewable ecological resources used up are not properly taken into account, or renewable resources are destroyed faster than they can regenerate, the entropic burdens fall on the ecological environment rather than on the economic system. But, as the economic system is necessarily coupled to and dependent on input from the ecosystem (see Fig. 14.3), the entropic burden in the latter will feedback on the economic system as diminished input, so the economic system becomes poorer as a result. In our model, poverty is absolute, as there is a finite optimum rate at which resources can be utilised and transformed. This also means that when individuals amass excessive resources and incomes, others become poorer in absolute terms. Poverty threatens the survival of the system as a whole. It represents an unbalanced, uncoupled, highly dissipative system (see below).

Evidence that money is both energetic and entropic is provided by the well-known unreliability of assessing the real wealth of nations - equivalent to the mobilisable energy or resources stored in our model - from the Gross National Product (GNP) - the amount of spending, in US dollars, on goods and services carried out by various sectors of the economy.

Korten[31] points out that the more environmentally burdensome ways of meeting a given need are generally those that contribute most to the GNP. For example, driving a car contributes more than riding a bicycle, turning on an air-conditioner adds infinitely more than opening a window, relying on processed foods contributes more than eating natural foods in reusable containers, and so on. According to Daly and Cobb,[32] if one reconstructs U.S. GNP for 1960 to 1986 counting only those increases in output that relate to improvements in wellbeing and adjusting for depletion of human and environmental resources, their Index of Economic Welfare reveals that individual welfare peaked in 1969, then remained on a plateau before falling during the early and mid-1980s. Yet, from 1969 to 1986, GNP per person went up by 35% and fossil fuel consumption increased by around 17%. "Instead of creating wealth, our money system is depleting our real wealth: our communities, ecosystems, and productive infrastructure."[33]

Another suggestive estimate of energetic yield versus entropic costs comes from the comparison of 25 rice cultivation systems[34] (see Table 1), of which 8 are *pre-industrial* in terms of low fossil fuel input (2-4%) and high labour input (35-78%); 10 are *semi-industrial* with moderate to high fossil fuel input (23 - 93%) and low to moderate labour input (4 - 46%); and 7 are *full-industrial* with 95% fossil fuel input and extremely low labour input of 0.04 to 0.2%. The total output per hectare, calculated in GigaJoule (GJ, unit of energy) in pre-industrial systems fall into low and high output subgroups, the output of the low subgroup is one-twentieth to one-fifth of the full industrial yield. However, *the output of the high subgroup is two to three times that of the full-industrial systems.* The yield of semi-industrial systems are more homogeneous, with an average of 51.75 GJ, while the yield of full-industrial systems, even more uniform, average 65.99GJ. When the ratio of total energetic output to total input is examined, however, pre-industrial systems range between 6.9 and 11.5, with the figures for the most productive systems being as high as 15.3 to 29.2. Semi-industrial systems give ratios of 2.1 to 9.7, whereas the ratios of full-industrial systems are not much better than unity.

These figures illustrate the law of diminishing returns remarkably well: there seems to be a plateau of output per hectare around 70-80 GJ, regardless of the total input, which is only exceeded in the 3 high yielding pre-industrial systems of Yunnan, China. Intensifying energy input leads to a drop in efficiency - the ratio of output to input - which is particularly sharp as input approaches the output ceiling. The drop in efficiency reflects the increasing entropic costs of high rates of dissipation that occur when the rate of energy input exceeds the capacity of the system to effectively store the energy input.

The exceptionally high output of the Chinese systems is also an indication that the energy-storing capacity, or carrying capacity of a system *can* increase, depending, in particular, on the space-time differentiation and the dynamic closures introduced. For example, mixed cropping and crop rotation would increase space-time differentiation, while the utilisation of farmyard and human manure as organic fertilisers, which has been traditionally practised in China, would increase dynamic closure.

Table 1. Balance sheet of energetic costs and yields of rice cultivation systems

Location	Input (I)	Fossil Fuel (GJ)	Labour (%)	Output (O) (%)	O/I (GJ)
Pre-industrial					
1.Dayak, Sarawak (1951)	0.30	2	44	2.4	8.0
2. Dayak, Sarawak (1951)	0.63	2	51	5.7	9.0
3. Kilombero, Tanzania (1967)	0.42	2	39	3.8	9.0
4. Kilombero, Tanzania (1967)	1.44	3	35	9.9	6.9
5. Iban, Sarawak (1951)	0.27	3	36	3.1	11.5
6. Luts'un, Yunnan (1938)	8.04	3	70	166.9	20.8
7. Yits'un, Yunnan (1938)	10.66	2	78	163.3	15.3
8. Yut'sun, Yunnan (1938)	5.12	4	53	149.3	29.2
Semi-industrial					
9. Mandya, Karnataka (1955)	3.33	23	46	23.8	7.1
10. Mandya, Karnataka (1975)	16.73	74	16	80.0	4.8
11. Philippines (1972)	12.37	86	5	39.9	3.2
12. Philippines (1972)	16.01	89	4	51.8	3.2
13. Japan (1963)	30.04	90	5	73.3	2.4
14. Hongkong (1971)	31.27	83	12	64.8	2.1
15. Philippines (1965)	3.61	93	13	25.0	6.9
16. Philippines (1979)	5.48	33	16	52.9	9.7
17. Philippines (1979)	6.90	80	11	52.9	7.7
18. Philippines (1979)	8.72	86	7	52.9	6.1
Full-industrial					
19. Surinam (1972)	45.9	95	0.2	53.7	1.2
20. USA (1974)	70.2	95	0.02	52.9	0.8
21. Sacramento (1977)	45.9	95	0.04	80.5	1.8
22. Grand Prairie (1977)	52.5	95	0.04	58.6	1.1
23. S.W. Louisiana (1977)	48.0	95	0.04	50.8	1.1
24. Mississippi delta (1977)	53.8	95	0.05	55.4	1.0
25. Texas Gulf Coast (1977)	55.1	95	0.04	74.7	1.4

The establishment of thriving microbial communities in the soil will also contribute to increased space-time differentiation for energy storage as well as improve dynamic closures by facilitating nutrient recycling (see below). By contrast, monocultures destroy space-time differentiation and dynamic closures both in the larger ecological community and in the microbial ecology of the soil, leading to diminished energy storage capacity, despite the high levels of input.

Dynamic Closure, Space-Time Differentiation and Cooperative Reciprocity

As emphasised earlier, a sustainable system has two branches, the irreversible, dissipative branch, and the non-dissipative branch, which requires dynamic closure or cycles, space-time differentiation and cooperative reciprocity. These principles have been well-appreciated by traditional indigenous farmers in their "internal input farming system" as Vandana Shiva[35] points out. The internal input farming system depends on a reciprocal, symbiotic relationship between individuals in the farming community, as well as between farmers and their 'produce'. As human beings farm, tend and propagate animals and plants, the animals and plants provide sustenance for the human community.[36] This involves the minimisation of waste by the judicious recycling of nutrients and diversification in the use of resources. By contrast, the so-called "high yielding varieties" introduced by the Green revolution are designed to destroy social relationships, break nutrient cycles and dispense with recycling, to depend on intensified external input. These are the reasons why fully industrialised intensive agricultural practices are so energy inefficient, and ultimately non-sustainable.

In the global context, the current proposals of the World Trade Organisation to remove all barriers to trade, finance and investment threaten not only the survival of local, national economies, but the survival of the global system as a whole.[37] The reason is that they destroy the nested levels of dynamic closures and space-time differentiation that enable the system to store energy and to operate in the cooperative reciprocity that contributes to the sustainability of the whole system. Without effective barriers which can guarantee the cooperative reciprocity on which local autonomy depends, the aggressive monopolies of multinational companies will further accelerate the extraction of cheap, natural resources while inflating the price of goods. Both those measures will lead to further increases in entropic costs, the one on the ecological environment, the other on the system itself, that will certainly spell the death of the global system.

Already, enlightened sectors in business are learning the lesson of sustainability from nature. On account of the dynamic closures involved in living cycles, there should really be no such thing as 'waste'. The most important species in the grand cycle of the earth's ecosystem are the scavengers that feed on wastes, carcasses and debris, thus releasing nutrients for green plants which capture the energy of sunlight to pass on to animals and other plants in the food-web. "Everything that is excreted, exhaled, or exhausted from one organism is used by another"[38] Following this principle to its logical conclusion, industrial processes are now being designed to turn waste products into valuable inputs.

For example, a new brewery in Namibia in southern Africa uses spent grain to grow mushrooms, then earthworms. The earthworms are used to feed chickens. The waste alkaline water from the brewery, which would have constituted 80% of the ground water extracted, is used first for cultivating *Spirulina*, an edible alga rich in protein, and then channelled to fish ponds where fish farming is introduced. Finally, the chicken manure goes through a digester to produce methane gas to be used for fuel. "This integrated biosystem will produce seven times more food, fuel, and fertiliser than a conventional operation and four times as many jobs."[39] This is none other than the minimisation of dissipation through increase in dynamic closure and space-time differentiation, so that the carrying capacity of the system is increased for a given rate of extraction of resources.

Conclusion

It is clear that we need a systematic change in direction in all spheres of life, before the dreams of solving all the problems facing the world today by genetic engineering biotechnology turn into nightmares. In this final chapter, I have given some indications of the sort of changes that are needed, and are already occurring. Fortunately for us, there is a substantial body of knowledge that goes beyond reductionist science to a deeper understanding of the organic whole. Some of this knowledge is already being put into practice in health, agriculture and industrial processes. Reductionist science has had its day. Let us reject the bad science that has served to exploit, to oppress, to obfuscate and to destroy the earth and its inhabitants. Let us opt for a joyful and sustainable future - beyond genetic engineering.

Notes

1. Soros, 1997, p.1

2. Soros, 1997, p.1.

3. Strohman (1997) describes a coming revolution in biology not too dissimilar to the one I am putting forth here. He points out how anomalies within the genetic determinist paradigm are being merely swept under the rug by "expert but conservative elements within the scientific mainstream", in order to rescue the paradigm. But the explanations [for behaviour of complex systems] become so contorted and invoke so many genetic agents and their "interactive and codependent states" that they are of questionable value.

4. See Ho, 1993; 1995a,b; 1996b; 1997a.

5. See Rubin, 1992, p. 2.

6. Ho, 1996b, p. 263.

7. Guyton, 1980, p.7.

8. Rubin, 1992, p.1.

9. Kennedy *et al*, 1980; Mondal and Heidelberger, 1970.

10. Brouty-Boyé *et al*, 1979; Rubin *et al*, 1990.

11. Rubin, 1992, p.1.

12. See Leape, *et al*, 1991. Iatrogenic diseases may even be heritable, as revealed by recent studies involving rats and rabbits that thalidomide-induced abnormalities may be passed on to the next generation. See "Thalidomide curse 'may go on down generations' " Phillip Knightley, *Independent on Sunday*, 20 April, p.3, 1997.

13. See literature produced by The Cancer Help Centre, Grove House, Cornwallis Grove, Clifton, Bristol BS8 4PG.

14. Costerton *et al*, 1995.

15. Freter, 1986.

16. "New and resurgent diseases. The failure of attempted eradication", *The Ecologist* 25, 21-26, 1995. See also Morse, 1993.

17. Edward Goldsmith started the magazine, *The Ecologist*, in 1970. Its famous manifesto, "A Blueprint for Survival", appeared as a special issue of *The Ecologist* in 1971.

18. Rifkin and Howard, 1980.

19. Daly, 1996.

20. Martin Khor, seminar on Globalization and Economics, Schumacher College, February, 1997.

21. Rifkin and Howard, 1980.

22. This theory of the organism has been developed over a number of years, see Ho, 1993; 1994; 1995a,b; 1996b,e; 1997a. The description here is necessarily a very brief summary.

23. Martin Khor, seminar on Globalization and Economics, Schumacher College, February, 1997.

24. Daly, 1996.

25. Lovelock, 1979; see also Saunders, 1994.

26. "Geophysiological Society: Mission Statement", Lee Kump, 1996.

27. Schneider and Kay, 1994.

28. Luvell and Holbo, 1991.

29. Ho, 1997d.

30. Daly, 1996.

31. Korten, 1995.

32. Daly and Cobb, 1989.

33. Korten, 1997, p.14.

34. See Shiva, 1993.

35. Shiva, 1993.

36. The reciprocal, symmetrical relationship between human beings and nature is beautifully brought out in Grimaldo Rengifo Vasquez's (1997) description of the Peruvian farming communities which practise a sustainable agriculture so intimately integrated with the ecosystem that there is no separation between nature and culture. Peasants collect seeds from differ-

ent places and grow them, adopting the plants as members of their family. In one of the oldest rituals celebrating the harvest of the new crop, last year's potato speaks to the new potato: as I bred these human beings, now I pass [this power] on to you. "The meaning of life," says Grimaldo, "is not only to breed, but to allow oneself to be bred."

37.See many excellent chapters arguing against all aspects of the global economy in Mander and Goldsmith, 1996.

38.Mshigeni and Pauli, 1997, p. 41.

39. Mshigeni and Pauli, 1997, p.42.

Moratorium on GE biotechnology and No to Patents on Life

During the 1997 State of the World Forum (Nov. 4-9) in San Francisco, Brian Goodwin, Fritjof Capra, Ervin Laszlo and myself took part in the Round Table on "New Visions in Biological Sciences". As part of our contribution to the new visions, we drafted a statement (reproduced on the next page) rejecting patents on life and calling for a moratorium on commercial release of genetic engineered organisms pending a comprehensive public enquiry. We plan to collect 100 000 signatures in the first instance to present to
1.The next State of the World Forum,
2. The UN Convention on Biological Diversity,
3. Our Government(s), and
4. The WHO, to request an enquiry into genetic engineering and the recent resurgence of infectious diseases. Please sign on, and collect more signatures by duplicating this page. and mail to: Mae-Wan Ho Biology Department, Open University, Walton Hall, Milton Keynes MK7 6AA, U.K. Alternatively, e-mail details to M.W. Ho <m.w.ho@open.ac.uk> stating that you agree to sign onto the Statement on Life and Evolution.

Among the signatories are:

Miguel A. Altieri, University of California, Berkeley, USA;
Frappé Benoît, Parti de la Loi Naturelle, Montlignon, France;
Fritjof Capra, Institute for Ecoliteracy, California, USA;
Ronnie Cummins, Pure Foods Campaign, USA;
Kristin Dawkins, and Yvette Flynn, Institute for Agriculture and Trade Policy, USA;
Kristin Ebbert, Mothers & Others for a Livable Planet / THE GREEN GUIDE, New York, USA;
Peter Fenwick and David Lorimer, Scientific and Medical Network, U.K.;
Brian Goodwin, Schumacher College, U.K.
Mae-Wan Ho, Open University, U.K.;
Vyvyan Howard, University of Liverpool, U.K.;
Tewolde Egziagher, and Sue Edwards, Institute of Sustainable Agriculture, Ethiopia,
Ervin Laszlo, Club of Budapest, Hungary;

Tore Midtvedt, Laboratory of Medical Microbial Ecology Cell & Molecular Biology, Stockholm, Sweden;

Bob Phelps, Australian GeneEthics Network, Fitzroy. Australia;

Darrell Posey, Oxford Centre for the Environment, Ethics & Society, Mansfield college, Oxford. UK;

Marilyn Schlitz, Institute of Noetic Sciences, California, USA;

Peter Saunders, King's College, London, U.K.;

Beatrix Tappeser, Institute for Applied Ecology, Freiburg, Germany; Brian Tokar, Inst. of Social Ecology, Vermont, USA;

Terje Traavik, Institute of Medical Microbiology, Tromso, Norway;

Peter R. Wills, University of Auckland, New Zealand.

1997 State of the World Forum Statement on Life and Evolution

Life is an intimate web of relations that evolves in its own right, interfacing and integrating its myriad diverse elements. The complexity and interdependence of all forms of life have the consequence that the process of evolution cannot be controlled, though it can be influenced. It involves an unpredictable creative unfolding that calls for sensitive participation from all the players, particularly from the youngest, most recent arrivals, human beings.

Life must not be treated as a commodity that can be owned, in whole or in part, by anyone, including those who wish to manipulate it in order to design new life forms for human convenience and profit. There should be no patents on organisms or their parts. We must also recognize the potential dangers of genetic engineering to health and biodiversity, and the ethical problems it poses for our responsibilities to life. We propose a moratorium on commercial releases of genetically engineered products and a comprehensive public enquiry into the legitimate and safe uses of genetic engineering. This enquiry should take account of the precautionary principle as a criterion of sensitive participation in living processes. Species should be respected for their intrinsic natures and valued for their unique qualities, on which the whole intricate network of life depends.

We recognise the validity of the different ways of knowing that have been developed in different cultures, and the equivalent value of the knowledge gained within these traditions. These add substantially to the set of alternative technologies that can be used for the sustainable use of natural resources that will allow us to preserve the diversity of species and to pass the precious gift of life in all its beauty and creativity to our children and their children, to the next century and beyond.

Name ____

Title ____

Affiliation (if any) __________________________________

Address ____

____

Signed ____

References

Abbott, A. (1996). Complexity limits the powers of prediction. *Nature* 379, 390.

Allison, R. (1995). RNA plant virus recombination. *Proceedings of USDA-APHIS/AIBS Workshop on Transgenic Virus-resistant Plants and New Plant Viruses*, April 20-21, Beltsville, Maryland.

Alper, J. (1996). Genetic complesity in single gene diseases. *British Medical J.* 312, 196-1971

Altieri, M.A. (1991). Traditional farming in Latin America. *The Ecologist* 21, 93-96.

Amabilecuevas, C.F. & Chicurel, M.E. (1993). Horizontal gene transfer. *American Scientist* 81, 332-341.

Anderson, E.S. (1975). Viability of, and transfer of a plasmid from E. coli K12 in the human intestine. *Nature* 255, 502.

Atlas, M., Bennett, A.M., Colwell, R., Van Elsas, J., Kjelleberg, S., Pedersen, J. & Wacker-Nagel, S. (1992). Persistence and survival of genetically-modified microorganisms released into the environment. In *The Release of Genetically Modified Microorganisms* (D. E. S. Stewart-Tull & M. Sussman eds.), p.117, Plenum Press, New York.

Avery, O.T., MacLeod, C.M. & McCarty, M. (1944). Studies on the chemical nature of the substance inducing transformation in pneumococcal types. *J. of Experimental Medicine* 79, 137-159.

Bains, W. (1987). *Genetic Engineering for Almost Everybody*, Penguin Books, London.

Baltimore, D. (1985). Retroviruses and retrotransposons: the role of reverse transcription in shaping the eukaryotic genome. *Cell* , 481-482.

Barinaga, M. (1996). A shared strategy for virulence. *Science* 272: 1261-1263.

Barzun, J. (1958). *Darwin, Marx and Wagner*, Doubleday Anchor, New York.

Bauer, L.S. (1995). Resistance - a threat to the insecticidal crystal proteins of *Bacillus-thuringiensis*. *Florida Entomologist* 78, 414-443.

Bhattacharyya, M.K., Smith, A.M., Ellis, Th.H.N., Hedley, C. & Martin, C. (1990). The wrinkeld-seed character of pea described by Mendel is caused by a transposon-like insertion in a gene encoding starch-braching enzyme. *Cell* 80, 115-122.

Belonga, E.A., Hedberg, C.W., Gleich, G.J., White, K.E., Mayeno, A.R., Loegering, D.A., Dunnette, S.L., Pirie, P.L., MacDonald, K.L. & Osterholm, M.T. (1990). An investigation of the cause of the eosinophilia-myalgia syndrome associated with tryptophan use. *The New England J. of Medicine.* 323, 347-365

Bergh, O., Borsheim, K.Y., Bratbak, G. & Heldal, M. (1989). High abundance of viruses found in aquatic environments. *Nature* 340, 467-468.

Bieth, E. & Darlix, J.-L. (1993). Characterization of a molecular clone of a highly infectious avian leukosis virus. *C.R. Aademie Science, Paris, Sciences de la vie* 316, 754-762.

Bik, E.M., Bunschoten, A.E., Gouw, R.D. & Mooi, F.R. (1995). Genesis of novel epidemic *vibrio-cholerae*-0139 strain-evidence for horizontal transfer of genes involved in polysaccharide synthesis. *Embo J.* 14: 209-216.

Binet, A. (1913). *Les Idées modernes sur les enfants*, Flammarion, Paris.

Blanc, V., Jordana, X., Litvak, S. & Araya, A. Control of gene expression by base deamination: The case of RNA editing in wheat mitochondria. (1996) 78, 511-517.

Bootsma, J.H., Vandijk, H. Verhoef, J., Fleer, A. & Mooi, F.R. (1996). Molecular characterization of the bro b-lactamase of *moraxella* (Branhamella) catarrhalis. *Antimicrobial Agents and Chemotherapy* 40: 966-972.

Bostock C.J. & Tyler-Smith, C. (1982). Changes to genomic DNA in methotrexate-resistant cells. In *Genome Evolution* (G.A. Dover & R.B. Flavell, eds.), pp. 69-94, Academic Press, London.

Boulanger, D., Crouch, A., Brochier, B., Bennett, M., Clement, J., Gaskell, R.M.,

Baxby, D. & Pastoret, P.P. (1996). Serological survey for orthopoxvirus infection of wild mammals in areas where a recombinant rabies virus is used to vaccinate foxes. *Veterinary Record* 138, 247-249.

Bowcock, A.M. (1993). Molecular-cloning of *BRCA1* - a gene for early-onset familial breast and ovarian cancer. *Breast Cancer Research and Treatment* 28, 121-135.

Brennan, T.A., Leape, L.L., & Laird, N.M. 1991). Incidence of adverse events and negligence in hospitalized patients - results of the Harvard Medical Practice Study. *New England J. of Medicine* 324, 370-376.

Brouty-Boye, D., Gresser, I. & Baldwin, C. (1979). Reversion of the transformed phenotype to the parental phenotype by subcutivation of X-ray-transformed C3H/10T1/1 cells at low cell density. *international J. of Cancer* 24, 253-260.

Brown, D.F.J., Farrington, M. & Warren, R.E. (1993). Imipene-resistant *Escherichia coli. Lancet* 342, 177.

Brown, P. & Kleiner, K. (1994). Patent row splits breast cancer researchers, *New Scientist* 24 September, 4.

Burrows, B. (1995). Scientists charge US agency with acting irresponsibly. *Third World Resurgence* 63, 2-3.

Burrows, B. (1996). Second thoughts about U.S. Patent #4,438,032. *Bulletin of Medical Ethics* 124, 11-14.

Cairns, J. (1978). *Cancer: Science and Society*, Freeman, San Francisco.

Cairns, J., Overbaugh, J. & Miller, S. (1988). The origin of mutants. *Nature* 335, 142-145.

Campbell, J.H., Lengyel J.A., & Langridge, J. (1973). Evolution of a second gene for b-galactosidase in *E. coli. Proceedings of the National Academy of Sciences (USA)* 70, 1841-1845.

Chee, Y.L. (1996). Concerted moves to undermine a strong biosafety agreement. *Third World Resurgence* 74, 8-10.

Clewell, D.B. ed. (1993). *Bacterial Conjugation*, Plenum Press, New York.

Coffey, T.J., Dowson, C.G., Daniels, M. & Spratt, B.G. (1995). Genetics and molecular-biology of b-lactam-resistant *pneumococci.* Microbial *Drug Resistance-Mechanisms Epidemiology and Disease* 1: 29-34.

Coghlan, A. (1995). Gene dream fades away. *New Scientist* 25 November, 14-15.

Colman, A. (1996). Production of proteins in the milk of transgenic livestock: problems, solutions and successes. *American J. of Clinical Nutrition* 63, 639S-645S.

Cooking, E.C. (1989). Plant cell and tissue culture. In *A Revolution in Biotechnology* (J.L Marx, ed.),pp. 119-129, Cambridge Univ. Press, Cambridge, New York.

Costerton, J.W., Lewandowski, Z., deBeer, D., Caldwell, D., Korber, D. & James, G. (1994). Biofilms, the customized microniche. *J. of Bacteriology* 176, 2137-2142.

Cox, C. (1995). Glyphosate, Part 2: Human exposure and ecological effects. *J. of Pesticide Reform.* Vol 15, No.4

Chen, I.S.Y., Mclaughlin, J., Gasson, J.C., Clark, S.C. & Golde, D.W. (1983). Molecular characterization of genome of a novel human T-cell leukaemia virus. *Nature* 305, 502-505.

Cleaver, J.E. & Kraemer, K.H. (1989). *The Metabolic Basis of Inherited Diseases*, McGraw-Hill, New York.

Clewell, D.B. & Flannagan, S.E. (1993). The conjugative transposons of gram-positive bacteria. In *Bacterial Conjugation* (D.B. Clewell, ed.), pp. 369-393, Plenum Press, New York.

Coghlan, A. (1996). Gene shuttle virus could damage the brain. *New Scientist* May 11, 6.

Cohen, J. (1996). New role for HIV: a vehicle for moving genes into cells. *Science* 272, 195.

Cohen, P. (1996). Doctor, there's a fly in my genome. *New Scientist* March 9, 6.

Collis, C.M., Grammaticopoulous, G., Briton, J., Stokes, H.W. and Hall, RM. (1993). Site-specific insertion of gene cassettes into integrons. *Molecular Microbiology* 9, 41-52.

Colman, A. (1996). Production of proteins in the milk of transgenic livestock -

problems, solutions and successes. *Am. J. Clin. Nutrition* 63:S639-S645.

Commandeur, P. & Komen, J. (1992). Biopesticides: Options for biological pest control increase. *Biotech Develop. Monitor* 13 (Dec): 6-7.

Costerton, J.W., Lewandowski, Z., DeBeer, D., Caldwell, D., Korber, D. & James, G. (1994). Biofilms, the customized microniche. *J. Bacteriol.* 176: 2137-2142.

Courvain, P., Goussard, S. & GrillotCourvain, C. (1995). Gene transfer from bacteria to mammalian cells. *Comptes Rendus de L'Academie Des Sciences Serie III-Science de La Vie*, 318, 1207-1212.

Creamer, R. & Falk, B.W. (1990). Direct detection of transcapsidated barley yellow dwarf luteoviruses in doubly infected plants. *J. Gen. Virol.* 71: 211-217.

Cropper, A. (1994). A novel approach. *Our Planet* 6, (back cover).

Cullis, C.A. (1983). Environentally induced DNA changes in plants. *CRC Critical Reviews in Plant Science* 1, 117-131.

Cullis, C.A. (1988). Control of variation in higher plants. In *Evolutionary Processes and Metaphors* (M.W. Ho & S.W. Fox, eds.), Wiley, London.

Cummins, J. (1994). The use of cauliflower mosaic virus. 35S promoter (CaMV) in Calgene's Flav Savr tomato creates hazard. Available from the author at <jcummins@julian.uwo.ca>

Cummins, J. (1997). Insecticide viruses for insect control. Available from the author at <jcummins@julian.uwo.ca>

Dalsgaard, K., Uttenthal, A., Jones, T.D., Xu, F., Merryweather, A., Hamilton, W.D.O., Langeveld, J.P.M., Boshuizen, R.X., Kamstrup, S., Lomonossoff, G.P., Porta, C., Vela, C., Casal, J.I., Eloen, R.H. & Rodgers, P.B. (1997). Plant-derived vaccine protects target animals against a viral disease. *Nature Biotechnology* 15, 248-252.

Daly, H. E. (1996). *Beyond Growth, The Economics of Sustainable Development*, Beacon Press, Boston.

Daly, H. E. & Cobb, J.B., Jr. (1989). *For the Common Good: Redirecting the Economy toward Community, the Environment and a Sustainable Future*, Beacon Press, Boston,

Damency (1994). The impact of hybrids between genetically modified crop plants and their related species: introgression and weediness. *Molecular Ecolology* 3: 37-40.

Dangler, C.A., Deaver, R.E. & Koloziej, D.M. (1994). Measurement of Aujeszkys-Disease virus recombination in-vitro under conditions of low multiplicity of infection. *Acta Veterinaria Hungarica* 42, 205-208.

Darwin, C. (1859). *The Origin of Species*, Murray, London.

Dawkins, R. (1976). *The Selfish Gene*, Oxford Univ. Press, Oxford.

Davies, J. (1994). Inactivation of antibiotics and the dissemination of resistance genes. *Science* 264: 375-382.

Denotter, W., Merchant, T.E., Beijerinck, D. & Koten, J.W. (1996). Breast-cancer induction due to mammographic screening in hereditarily affected women. *Anticancer Research* 16, 3173-3508.

DeAngelis, D.L. (1992). *Dynamics of Nutrient Cycling and Food Webs*, Chapman and Hall, London.

DiFronzo, N.L. & Holland, C.A. (1993). A direct demonstration of recombination between an injected virus and endogenous viral sequences, resulting in the generation of mink cell focus-inducing viruses in AKR mice. *J. of Virology* 67, 3763-3770, 1993.

Doerfler, W. (1991). Patterns of DNA Methylation - evolutionary vestiges of foreign DNA inactivation as a host defense mechanism. *Biol. Chem. Hoppe-Seyler* 372: 557-564.

Doerfler, W. (1992). DNA methylation: eukaryotic defense against the transcription of foreign genes? *Microbial Pathogenesis* 12: 1-8.

Doucet-Populaire, F. (1992). Conjugal transfer of genetic information in gnotobiotic mice, In *Microbial Releases* (M.J. Gauthier, ed.), Springer Verlag, Berlin,

Dover, G. A. & Flavell, Ed. (1982). *Genome Evolution*, Academic Press, London.

Eber, G., Chevre, A.M. Baranger, A., Vallee, P., Tanfuy, X. & Renard, M.

(1994). Spontaneous hybridization between a male-sterile oilseed rape and two weeds. *Theor. App. Gene.* 88: 362-368.

Egziabher, T.B.C. (1994). Where is the GOOD WILL? *Our Planet* 6, 17-19.

Elnatan, J., Goh, H.S. and Smith, D.R. (1996). C-KI-RAS activation and the biological behavior of proximal and distal colonic adenocarcinomas. *European J. of Cancer* 32A, 491-497.

Estruch, J.J., Carozzi, N.B., Desai, N., Duck, N.B., Warren, G.W. & Doziel, M.G. (1997). Transgenic plants: An emerging approach to pest control. *Nature Biotechnology* 15, 137-141.

Fernando, S. (1990). The same difference. *New Internationalist*, issue on Madness, July, 24-25.

Finnegan H. & McELroy (1994). Transgene inactivation plants fight back! *Bio/Techology* 12: 883-888.

Flavell, R.B. (1982). Sequence amplification, deletion and rearrangement: major sources of variation during species divergence. In *Genome Evolution* (G.A. Dover & R.B. Flavell, eds.) pp. 301-323, Academic Press, London.

Foster, P.L. (1992). Directed mutation: between unicorns and goats. *J. of Bacteriology* 174, 1711-1716.

Frank, S. & Keller, B. (1995). Produktesicherheit von krankheitsresistenten Nutzpflanzen: Toxikologie, allergenes Potential, Sekundäreffekte und Markergene Eidg. Forschungsantalt für landwirtschaftlichen Pflanzenbau, Zürich.

Franke, A.E. & Clewell, D.B. (1981). Evidence for a chromosome-borne resistance transposon (Tn*916*) in *Streptococcus faecalis* that is capable of 'conjugal' transfer in the absence of a conjugative plasmid. *J. of Bacteriology* 145, 494-502.

Freter, R. (1986). The need for mathematical models in understanding colonization ;and plasmid transfers in the mammalian intestine. In *Bacterial Conjugation* (D.B. Clewell, ed.), pp. 81-93, Plenum Press, New York.

Friedman, L.S., Gayther, A., Kurosaki, T., Gordon, D., Noble, B., Casey, G., Ponder, B.A.J. & Anton Culver, H. (1997). Mutation analysis of BRCA1 and BRCA2 in a male cancer population. *American J. of Human Genetics* 60, 313-319.

Frischer, M.E., Stewart, G.J. & Paul, J.H. (1994). Plasmid transfer to indigenous marine bacterial-populations. *FEMS Microbiol. Ecol.* 15: 127-135.

Garrett, L. (1995). *The Coming Plague Newly Emerging Disease in a World Out of Balance*, Penguin Books, New York.

George, S. (1988). *A Fate Worse Than Debt*, Penguin, Harmondsworth.

Gierl, A. (1990). How maize transposable elements escape negative selection. *Trends in Genetics* 6, 155-158.

Goldsmith, E. (1992). Development: fictions and facts. *Ecoscript* 35, Foundation for Ecodevelopment, Amsterdam.

Goldsmith, E. & Hildyard, N. (1991). World agriculture: Toward 2000, FAO's plan to feed the world. *The Ecologist* 21, 81-92.

Golovkina, T.V., Jaffe, A.B. & Ross, S.R. (1994). Coexpression of exogenous and endogenous mouse mammary tumor virus RNA in vivo results in viral recombination and broadens the virus host range. *Jour. of Virology* 68, 5019-5026.

Goodman, A.E., Marshall, K.C. & Hermansson, M. (1994). Gene transfer among bacteria under conditions of nutrient depletion in simulated and natural aquatic environments. *FEMS Microbiology Ecology* 15, 55-60.

Goodwin, B.C. (1984). A relational or field theory of reproduction and its evolutionary implications. In *Beyond neo-Darwinism: Introduction to the New Evolutionary Paradigm* (M.W. Ho and P.T. Saunders, eds.), pp. 219-241, Academic Press, London.

Goodwin, B.C. (1994). *How the Leopard Changed Its Spots: The Evolution of Complexity*, Weidenfeld and Nicolson, London.

Green, A.E. & Allison, R.F. (1994). Recombination between viral RNA and transgenic plant transcripts.*Science* 263: 1423.

Greenstein, J.L. & Sachs, D.H. (1997). The use of tolerance for transplantation across xenogeneic barriers. *Nature Bio-*

technology 15, 235-237.

Gu, Z., Gao, Q., Fast, E.A. & Wainberg, M.A. (1995). Possible involvement of cell fusion and viral recombination in generation of human immunodeficiency virus variants that display dual resistance to AZT and 3TC. *J. of General Virology* 76, 2601-2605.

Gudkov, A. & Kopnin, B. (1985). Gene amplification in multidrug-resistant cells: molecular and karyotypic events. *BioEssays* 3, 68-71.

Guillot, J.F. & Boucaud, J.L. (1992). *In vivo* transfer of a conjugative plasmid between isogenic Escherichia coli strains in the gut of chickens in the presence and absence of selective pressure In *Microbial Releases* (M.J. Gauthier, ed.) , pp. 167-174, Springer Verlag, Berlin.

Gustafson, T.L., Lievens, A.W., Brunell, P.A., Moellenberg, R.G., Buttery, C.M.G. & Sehulster, L.M. (1987). Measles outbreak in a fully immunized secondry-school population. *The New England J. of Medicine* 316, 771-774.

Guyton, A.C. *Physiology*, W.B. Saunders Company, Philadelphia.

Gurdon, J.B. (1974). *The Control of Gene Expression in Animal Development*, Oxford Univ. Press, Oxford.

Hall, B.G. & Hartl, D.L. (1974). Regulation of newly evolved enzymes I. Selection of a noel lactase regulated by lactose in *Escherichia coli. Genetics* 76, 391-400.

Hama, H., Suzuki, K. & Tanaka, H. (1992). Inheritance and stability of resistance to *Bacillus thuringiensis* formulations in diamondback moth, *Plutella xylostella* (Linnaeus) (Lepidoptera: Yponomeutidae). *Appl. Entomol. Zool.* 27: 355-362.

Hamilton, J. O'C. & Carey, J. (1994). Biotech. An industry crowded with players faces an ugly reckoning. *Business Week* September 26, 66-72.

Harding, K. (1996). The potential for horizontal gene transfer within the environment. *Agro-Food-Industry Hi-Tech* July/August, 31-35.

Hardy, R.W.F. (1994). Current and next generation agricuoltural biotechnology products and processes considered from a public good perspective. *NABC Report 6 "Agricultural Biotechnology and the Public Good"* (J.F. MacDonald ed.), pp.43-50.

Harriman, E. (1988). The good old British jab. New *Statesman and Society* 12 Septermber, 10-12.

Heaton, M.P. & Handwerger, S. (1995). Conjugative mobilization of a vancomycin resistance plasmid by a putative enterococcus-faecium sex-pheromone response plasmid. *Microbial Drug Resistance-Mechanisms Epidemiology and Disease* 1: 177-183.

Heitman, D. & Lopes-Pila, J.M. (1993). Frequency and conditions of spontaneous plasmid transfer from *E. coli* to cultured mammalian cells. *BiosSystems* 29: 37-48.

Hermansson, M. & Linberg, C. (1994). Gene transfer in the marine environment. *FEMS Microbiology Ecology* 15: 47-54.

Hildyard, N. (1991). An open letter to Edouard Saouma, Director-General of the Food and Agriculture Organization of the United Nations. *The Ecologist* 21, 43-46.

Hildyard, N. (1996). Too many for what? The social generation of food "scarcity" and "overpopulation". *The Ecologist* 26, 282-289.

Ho, M.-W. (1976). *Molecular Genetics, S299 Genetics A Second Level Open Univ. Course*, Open Univ. Press, Milton Keynes

Ho, M.-W. (1984a). Environment and heredity in development and evolution. In *Beyond neo-Darwinism: Introduction to the New Evolutionary Paradigm*, (M.-W. Ho and P.T. Saunders, eds.), pp. 267-289, Academic Press, London.

Ho, M.-W. (1984b). Where does biological form come from? *Rivista di Biologia* 77, 147-179.

Ho, M.-W. (1987a). Evolution by process, not by consequence: implications of the new molecular genetics for development and evolution. *Int. J. comp. Psychol.* 1: 3-27.

Ho, M.-W. (1987b). *Regulation of Gene Expression in Eukaryotes, S298 Genetics*, Open Univ. Press, Milton Keynes.

Ho, M.-W. (1987c). *Genes in Populations,*

S298 Genetics, Open Univ. Press, Milton Keynes.

Ho, M.-W. (1988a). Evolution by process, not by consequence In *Evolutionary Processes and Metaphors* (M.-W. Ho and S.W. Fox, eds.), pp. 117-144, John Wiley & Sons, London.

Ho, M.-W. (1988b). How rational can rational morphology be? *Rivista di Biologia* 81, 11-55.

Ho, M.-W. (1988c). Genetic fitness and natural selection: myth or metaphor. In *Evolution of Social Behavior and Integrative Levels* (G. Greenberg, and E. Tobach, eds.), 85-111, Lawrence Erlbaum, New Jersey.

Ho, M.-W. (1990). An exercise in rational taxonomy. *J. of Theoretical Biology* 147, 43-57.

Ho, M.-W. (1992). Development, rational taxonomy and systematics. *Rivista di Biologia* 85, 193-211.

Ho, M.-W. (1993). *The Rainbow and The Worm, The Physics of Organisms*, World Scientific, Singapore.

Ho, M.W. (1994). What is Schrödinger's negentropy? *Modern Trends in BioThermoKinetics* 3, 50-61.

Ho, M.-W. ed. (1995a). *Bioenergetics, S327 Living Process*, Open Univ. Press, Milton Keynes.

Ho, M.-W. (1995b). Bioenergetics and the coherence of organisms. *Neural Network World* 5, 733-750.

Ho, M.-W. (1996a). Are current transgenic technologies safe? Capacity building in biosafety urgently needed for developed countries. In *Biosafety Capacity Building: Evaluation Criteria Development* (I. Virgin and R.J. Frederick, eds.), pp.75-80, Stockholm Environment Institute, Stockholm.

Ho, M.-W. (1996b). The biology of free will. *J. of Consciousness Studies* 3, 231-244.

Ho, M.-W. (1996c). Natural being and coherent society. pp 286-307 in *Gaia in Action, Science of the Living Earth* (Ed. P. Bunyard), Floris Books, Edinburgh,

Ho, M.-W. (1996d). Why Lamarck won't go away. *Ann. Human Genetics* 60: 81-84.

Ho, M.-W. (1996e). Bioenergetics and biocommunication. In *Computation in Cellular and Molecular Biological Systems* (R. Cuthbertson, M. Holcombe, and R. Paton eds.) pp. 251-262, World Scientific, Singapore.

Ho, M.-W. (1997a). Towards a theory of the organism. *Integrative Physiology and Behavioural Research* (in press).

Ho, M.-W. (1997b). Evolution. In *Handbook of Comparative Psychology* (G. Greenberg, and M. Haraway eds.), Garland Publishing, New York.

Ho, M.-W. (1997c) DNA and the new organicism. In *The Future of DNA* (J. Wirz and E. Lammerts van Bueren eds.) pp. 71-93, Kluwer Academic Publishers, Dordrecht.

Ho, M.-W. (1997d). On the nature of sustainable economic systems. *World Futures* (in press).

Ho, M.-W. & Fox, S.W. eds. (1988). *Evolutionary Processes and Metaphors*, Wiley, London.

Ho, M.-W. & Goodwin, B.C. (1987). *The Process of Heredity, Genetics S298*, Open Univ. Press, Milton Keynes.

Ho, M.-W., Popp. F.A. & Warnke, U. eds. (1994). *Bioelectrodynamics and Biocommunication*, World Scientific, Singapore.

Ho, M.-W. & Saunders, P.T. (1979). Beyond neo-Darwinism: an epigenetic approach to evolution. *J. of Theoretical Biology* 78, 573-591.

Ho, M.-W. & Saunders, P.T. eds. (1984). *Beyond neo-Darwinism: Introduction to the New Evolutionary Paradigm*, Academic Press, London.

Ho, M.-W. & Saunders, P.T. (1993). Rational taxonomy and the natural system - with particular reference to segmentation. *Acta Biotheoretica* 41, 298-304.

Ho, M.-W. & Saunders, P.T. (1994). Rational taxonomy and the natural system-segmentation and phyllotaxis. In *Models in Phylogeny Reconstruction* (R.W. Scotland, D.J. Siebert and D.M. Williams, eds.), pp. 113-124, Oxford Science, Oxford.

Ho, M.-W. & Tappeser, B. (1997). Potential contributions of horizontal gene transfer to the transboundary movement of living modified organisms resulting from modern biotechnology. *Proceedings of Work-*

shop on Transboundary Movement of Living Modified Organisms resulting from Modern biotechnology: Issues and Opportunities for Policy-makers (K.J. Mulongoy, ed.), pp. 171-193, International Academy of the Environment, Geneva.

Ho, M.-W., Tucker, C., Keeley, D. & Saunders, P.T. (1983). Effects of successive generations of ether treatment on penetrance and expression of the bithorax phenocopy in *Drosophila melanogaster. J. of Experimental Zoology* 225, 357-368.

Höfle, M.G. (1994). Auswirkungen der Freisetzung bakterieller Monokulturen auf die naturliche Mikroflora aquatischer Okosysteme. pp. 795-820 in *Biologische Sicherheit/Forschung Biotechnologie BMFT* (Ed Germany) vol.3, 1003pp.

Hoffman, T., Golz, C. & Schieder, O. (1994). Foreign DNA sequences are received by a wild-type strain of Aspergillus niger after co-culture with transgenic higher plants. *Current Genetics* 27: 70-76.

Holmes, T.M. & Ingham, E.R. (1995). The effects of genetically engineered microorganisms on soil foodwebs. In *Supplement to Bulletin of Ecological Society of America 75/2, Abstracts of the 79th Annual ESA Meeting: Science and Public Policy,* Knoxville, Tennessee, 2-77 August 1994.

Horrobin, D.F., Glen, A.I.M. & Hudson, C.J. (1995). Possible relevance of phospholipid abnormalities and genetic interactions in psychiatric-disorders - the relationship between dyslexia and schizophrenia. *Medical Hypotheses* 45, 605-613.

Hubbard, R. (1995). Genomania and health. *American Scientist* 83, 8-10.

Hubbard, R. & Wald, E. (1993) *Exploding the Gene Myth,* Beacon Press, Boston.

Hughes, V.M. & Datta, N. (1983). Conjugative plasmids in bacteria of the 'pre-antibiotic' era. *Nature* 302, 725-726.

Hyrien, O. & Buttin, G. (1986). Gene amplification in pesticide-resistant insects. *Trends in Genetics* 2, 275-276.

Inose, T. & Murata, K. (1995). Enhanced accumulation of toxic compounds in yeast cells having high glycolytic activity: a case study on the safety of genetically engineered yeast. *International J. of Food Science and Technology* 30, 141-146.

Ippen-Ihler, K. & Skurray, R.A. (1993). Genetic organization of transfer-related determinants on the sex factor F and related plasmids. In *Bacterial Conjugation* (D.B. Clewell, ed.), pp. 23-52, Plenum Press, New York.

Jablonka, E. & Lamb, M. (1995). *Epigenetic Inheritance and Evolution. The Lamarckian Dimension,* Oxford Univ. Press, Oxford.

Jacks, T., Fazeli, A., Schmitt, E.M., Bronson, R.T., Goodell, M.A. & Weinberg, R.A. (1992). Effects of an Rb mutation in the mouse. *Nature* 359, 295-299.

Jager, M.J. & Tappeser, B. (1995). Risk Assessment and Scientific Knowledge. Current data relating to the survival of GMOs and the persistence of their nucleic acids: Is a new debate on safeguards in genetic engineering required? - considerations from an ecological point of view. Preprint circulated and presented at the TWN-Workshop on Biosafety, April 10, New York.

Jeffreys, A.J., Wilson V. & Thein, S.L. (1985). Hypervariable 'minisatellite' regions in human DNA. *Nature* 314, 67-73.

Ji, H.J., Liu, Y.L.E., Wang, M.S., Liu, J.W. Xiao, G.W., Joseph, B.D., Rosen, C. & Shi, Y.E. (1997). Identification of a breast cancer-specific gene, BCSG1, by direct differential cDNA sequencing. *Cancer Research* 57, 759-764.

Johnston, A.W.B.. (1989). Biological nitrogen fixation. In *A Revolution in Biotechnology* (J.L. Marx, ed.), pp.103-118, Cambridge Univ. Press, Cambridge, New York.

Jones, S. & Taylor, K. eds. (1995). *Processes of Heredity, S327 Living Processes Book 4,* Open Univ., Milton Keynes.

Jones, S., Martin, R. & Pilbeam, D. (eds.) (1994). *The Cambridge Encyclopedia of Human Evolution,* Cambridge Univ. Press, Cambridge.

Jorgensen, R.B. & Andersen, B. (1994). Spontaneous hybridization between oilseed rape (*Brassica napus*) and weedy *B. campestris* (Brassicaceae): a risk of growing genetically modified oilseed rape. *Amer. Jour. Botany* 12: 1620-1626.

Kado, C.I. (1993). Agrobacterium-mediated transfer and stable incorporation of foreign genes in plants. In *Bacterial Conjugation* (D.B. Clewell, ed.), pp. 243-254, Plenum Press, New York.

Kapur, V., Kanjilal, S., Hamrick, M.R., Li, L.L., Whittam, T.A., Sawyer, S.A. & Musser, J.M. (1995). Molecular population genetic-analysis of the streptokinase gene of *Streptococcus-pyogenes*- mosaic alleles generated by recombination. *Molecular Microbiology* 16,: 509-519.

Kehoe, M.A., Kapur, V., Whatmore, A., & Musser, J.M. (1996). Horizontal gene transfer among group A streptococci: implications for pathogenesis and epidemiology. *Trends in Microbiology* 4, 436-443.

Kell, C.M., Hordens, J.Z., Daniels, M., Coffey, T.J., Bates, J., Paul, J., Gilks, C. & Spratt, B.G. (1993). Molecular epidemiology of penicillin-resistant *pneumococci* isolated in Nairobi, Kenya. *Infection and Immunity* 61: 4382-4391.

Kendrew, J., ed. (1995). *The Encyclopedia of Molecular Biology*, Blackwell Science, Oxford.

Kennedy, A.R., Fox, M., Murphy, G. and Little, J.B. (1980). Relationship between X-ray exposure and malignant transformation in C3H 10T1/2 cells. *Proceedings of the National Academy of Sciences USA* 77, 7262-7266.

Kerem, B., Rommens, J.M., Buchanan, J.A., Markiewica, D., Cox, T.K., Chakravarte, A., Buchwald, M. & Tsui, L.-C. (1989). Idenfication of the cystic fibrosis gene; genetic analysis. *Science* 245, 1073-1080.

Kimbrell, A. (1993). *The Human Body Shop. The Engineering and Marketing of Life*, The Third World Network, Penang.

King, J.L. & Briggs, R. (1955). Changes in the nuclei of differentiating gastrula cells, as demonstrated by nuclear transplantation. *Proceedings of the National Academy of Sciences, U.S. A.* 41, 321-325.

Kothari, A. (1994). The need for a protocol on farmers' rights and indigenous peoples. *Our Planet* 6, 39-40.

Korton, D.C. (1995). *When Corporations Rule the World*, Kumarian Press, West Hartford, Conn.

Korten, D.C. (1997). Civil societies versus the global economy: a struggle for life. *Presentation to the 22nd World Conference of the Society for Internation Development, Which Globalization? Opening Spaces for Civic Engagement, Plenary 1 on "Globalization and the Transformation of Political Debates"*. Text circulated by the author at the Conference.

Kropotkin, P. (1914). *Mutual Aid: A Factor of Evolution*, Extending Horizon Books, Boston, Mass.

Lamarck, J.B. (1809). *Philosophie Zoologique*, Paris.

Lau, P.P., Zhu, H.J., Nakamuta, M. & Chan, L. (1997). Cloning of an Apobec-1-binding protein that also interactx with apolipoprotein B mRNA and evidence for its involvement in RNA editing. *The J. of Biological Chemistry* 272, 1452-1455.

Laszlo, E. (1995). T*he Interconnected Universe*, World Scientific, Singapore.

Laszlo, E. (1996). *The Whispering Pond*, Element, Rockport, Mass.

Leape, L.L., Brennan, T.A., Lair, N., Lawthers, A.G., Localio, A.R., Barnes, B.A., Hebert, L., Newhouse, J.P., Weiler, P.C. & Hiatt, H. (1991). The nature of adverse events in hospitalized patients. Results of the Harvard medical practice study II. *The New England J. of Medicine* 324, 377-384.

Lebaron, Ph., Batailler, N. & Baleux, B. (1994). Mobilization of a recombinant nonconjugative plasmid at the interface between wastewater and the marine coastal environment. *FEMS Microbiology Ecology* 15, 61-70.

Lee, E.Y.H.P., Chang, C.Y., Hu, N., Wang, Y.C.J., Lai, C.C., Herrup, K. & Lee, W.H. (1992). Mice deficient for Rb are non-viable and show defects in neurogenesis and haematopoiesis. *Nature* 359, 288-294.

Lee, H.S., Kim, S.W., Lee, K.W., Ericksson, T. & Liu, J.R. (1995). Agrobacterium-mediated transformation of ginseng *(Panax-ginseng)* and mitotic stability of the inserted beta-glucuronidase gene in regenerants from isolated protoplasts.

Plant Cell Reports 14: 545-549.

Lemke, P.A. & Taylor, S.L. (1994). Allergic reactions and food intolerances. In *Nutritional Toxicology* (F.N. Kotsonis, M. Mackay & J.J. Hjelle, eds.), pp. 117-137, Raven Press, New York.

Lenoir, G., Narod, S., Olopade, O., Plummer, S., Ponder, B., Serova, O., Simar, J., Stratton, M. & Warren, B. (1996). Mutations and polymorphiss in the familial early-onset breast-cancer (BRCA1) gene. *Human Mutation* 8;, 8-18.

Leong, A.S.Y., Robbins, P., & Spagnolo, D.V. (1995). Tumor genes and their proteins in cytologic and surgical specimens - relevance and detection systems. *Diagnostic Cytopathology* 13, 411-422.

Levidow, L, Carr, S., von Schomberg, R. & Wield, D. (1996). Bounding the risk assessment of a herbicide-tolerant crop. In *Coping with Deliberate Release: The Limits of Risk Assessment* (A. van Doommelen, ed.), pp. 81-102, International Centre for Human and Public Affairs, Tilburg,

Levy, S.B. (1984). Playing antibiotic pool: Tie to tally the score. *New England J. of Medicine* 311, 663-664.

Levy, S.B. & Novick, R.P. eds. (1986). *Antibiotic Resistance Genes: Ecology, Transfer, and Expression*, 24 Banbury Report, Cold Spring Harbour Laboratory, New York.

Lewis, D.L. & Gattie, D.K. (1991). The ecology of quiescent microbes. *ASM News* 57: 27-32.

Lewontin, R.C. (1982). *Human Diversity*, Scientific American Books, W.H. Freeman, New York.

Lewontin, R. C. (1985). Whatever happened to eugenics? *GeneWatch* Jan. - Apr., 8-10.

Lin, S., Gaiano, N., Culp, P., Burns, J.C., Friedmann, T., Yee, J.-K. & Hopkins, N. (1994). Integration and germ-line transmission of a pseudotyped retroviral vector in zebrafish. *Science* 265: 666-669.

Longerich, S., Galloway, A.M., Harris, R.S., Wong, C. & Rosenberg, S.M. ((1995). Adaptive mutation sequences reproduced by mismatch repair deficiency. *Proceedings of the National Academy of Sciences (USA)* 92, 12017-12020,

Lorenz, M.G. & Wackernagel, W. (1994). Bacterial gene transfer by natural genetic transformation in the environment.*Microbiological Reviews* 58: 563-602.

Lovelock , J. (1979). *A New Look at Gaia*, Oxford Univ. Press, Oxford.

Lovelock, J. (1996). The Gaia hypothesis. In *Gaia in Action, Science of the Living Earth* (P. Bunyard, ed.), pp. 15-33, Floris Books, Edinburgh.

Lundstrom, K. & Turpin, M.P. (1996). Proposed schizophrenia-related gene polymorphism - expresssion of the ser9gly mutant human dopamine D-3 receptor with the semliki-forest-virus system. *Biochemical and Biophysical Research Communications* 225, 1068-1072.

Luvall, J.C. & Holbo, H.R. (1991). Thermal remote sensing methods in landscape ecology. In *Quantitative Methods in Landscape Ecology*, (M. Turner & R.H. Gardner, eds.) Chapter 6, Springer-Verlag, New York.

Maier, R.M., Zeltz, P., Kossel, H., Bonnard, G., Gualberto, J.M. & Grienenberger, J.M. (1996). RNA editing in plant mitochondria and chloroplasts. *Plant Molecular Biology* 32, 343-365.

Manavathu, E.K., Hiratsuka, K. & Taylor, D.E. (1988). Nucleotide sequence analysis and expression of a tetracycline resistance gene from *Campylobcter jejuni. Gene* 62, 17-26.

Mander, J. & Goldsmith, E. ed. (1996). *The Case Against the Global Economy and For a Turn Toward the Local*, Sierra Club Books, San Francisco.

Marshall, R. J. (1984). The genetics of schizophrenia revisited. *Bulletin of the British Psychological Society* 37, 177-181.

Mayeno, A.N. & Gleich, G.J. (1994). Eosinophilia-myalgia syndrome and tryptophan production: a cautionary tale. *Tibtech* 12, 346-352.

Mazodier, P. & Davies, J. (1991). Gene transfer between distantly related bacteria. *Annual Review of Genetics* 25: 147-171.

McClintock, B. (1984). The significance of

responses of the genome to challenge. *Science* 226, 792-801.

McNally, R. (1995). Genetic madness. The European rabies eradication programme. *The Ecologist* 24, 207-212.

McNally, R. & Wheale, P. (1996). Biopatenting and biodiversity. Comparative advantages in the new global order. *The Ecologist* 26, 5, 222-228.

Meister, I. & Mayer, S. (1994). *Genetically engineered plants: releases and impacts on less developed countries,* A Greenpeace inventory, Greenpeace International.

Mellon, M. & Rissler, J. (1995). Transgenic Crops:USDA data on small-scale tests contribute little to commercial risk assessment. *Bio/Technology* 13, 96.

Mezrioui, N. & Echab, K. (1995). Drug resistance in Salmonella strains isolated from domestic wastewater before and after treatment in stabilization ponds in an arid region (Marrakech, Morocco). *World J. of Microbiology & Biotechnology* 11, 287-290.

Mihill, C. (1996). Killer diseases making a comeback, says WHO. *Guardian* 10/5/96 p. 3.

Miller, H. (1995) "Don't need, don't look", Letter to the Editor, *Bio/Technology* 13, 201.

Mikkelsen, T.R., Andersen, B. & Jorgensen, R.B. (1996). The risk of crop transgene spread. *Nature* 380: 31.

Moens, L., Vanfleteren, J., Vandepeer, Y., Peeters, K., Kapp, O., Czeluzniak, J., Goodman, M., Blaxter, M. & Vinogradov, S. (1996). Globins in nonvertebrate species - dispersal by horizontal gene transfer and evolutio of the struture-function relationships. *Molecular Biology and Evolution* 13, 324-333.

Moffat, A. S. (1996). Biodiversity is a boon to ecosystms, not species. *Science* 271, 1497.

Mondal, S., & Heidelberger, C. (1970). In vitro malignant transformation by methylcholanthrene of the progeny of single cells derived from C3H mouse prostate. *Proceedings of the national Academy of Sciences USA* 65, 219-229.

Moore, J.A. (1955). Abnormal combinations of nuclear and cytoplasmic systems in frogs and toads. *Advances in Genetics* 7, 139-182.

Morgan, T.H. (1916). *A Critique of the Theory of Evolution,* Princeton Univ. Press, Princeton, pp. 187-190.

Morse, S.S. ed. (1993). *Emerging Virses,* Oxford Univ. Press, New York.

Mshigeni, K. &Pauli, G. (1997). Brewing and future, *Yes! A J. of Positive Futures* Spring, 41-43.

Mulvihill, J.J. (1995). Craniofacial syndromes: no such thing as a single gene disease. *Nature Genetics* 9, 101-103.

Neilson, J.W., Josephson, K.L., Pepper, I.L., Arnold, R.B., Digiovanni, G.D. & Sinclair, N.A. (1994). Frequency of horizontal gene-transfer of a large catabolic plasmid (PJP4) in soil. *App. Environ. Microbiol.* 60, 4053-4058.

Nijar, G.S. & Chee, Y.L. (1994). Intellectual property rights: the threat to farmers and biodiversity. *Third World Resurgence* 39, 8-10.

Nikaido, H. (1994). Prevention of drug access to bacterial targets: permeability barriers and active efflux. *Science* 264, 382-388.

Nordlee, J.A., Taylor, S.L., Townsend, JA., Thomas, L.A. & Bush, R.K. (1996). Identification of a brazil-nut allergen in transgenic soybeans. *The New England J. of Medicine* March 14, 688-728.

Offit, K., Gilewski, T. McGuire, P., Schluger, A., Hampel, H., Brown, K., Swensen, J., Neuhausen, S., Skolnick, M., Norton, L. & Goldgar, D. (1996). Germline BRCA1 185delAG mutations in Jewish women with breast-cancer. *Lancet,* 347, 1643-1645.

Olby, R.C. (1966). *Origins of Mendelism,* Constable, London.

Oldroy, D.R. (1980). *Darwinian Impacts,* Open Univ. Press, Milton Keynes.

Osbourn, J.K., Sarkar, S. & Wilson, M.A. (1990). Complementation of coat protein-defective TMV mutants in transgenic tobacco plants expressing TMV coat protein. *Virology* 179: 921-925.

Pain, S. (1997). The plague dogs. *New Scientist* 19 April, 32-37, 1997.

Palca, J. (1986). Living outside regulation.

Nature 324, 202.

Pang, Y., Brown, B.A., Steingrube, V.A., Wallance R.J., Jr., & Roberts, M.C. (1994). Tetracycline resistance determinants in *Mycobaterium* and *Streptomyces* Species. *Antimicrobial Agents and Chemotherapy.* 38, 1408-1412.

Paulkaitis, P. & Roossinck, MJ. (1996). Spontaneous change of a benign satellite RNA of cucumber mosaic virus to a pathogenic variant. *Nature Biotechnology* 14, 1264-1268.

Perlas, N. (1994). *Overcoming Illusions About Biotechnology*, Zed Books, London and Third World Network, Penang.

Perlas, N. (1995). Dangerous trends in agricultural biotechnology *Third World Resurgence* 38, 15-16.

Pimm, S.L. (1991). *Balance of Nature - Ecological Issues in the Conservation of Species and Communities*, The Univ. of Chicago Univ. Press, Chicago.

Pollard, J.W., ed. (1984). *Evolution Theory - Paths into the Future*, Wiley, London.

Pollard, J. W. (1984). Is Weismann's barrier absolute? In *Beyond neo-Darwinism: Introduction to the New Evolutionary Paradigm* (M.W. Ho and P.T. Saunders, eds,), pp 291-315, Academic Press, London.

Pollard, J. W. (1988). The fluid genome and evolution. In *Evolutionary Processes and Metaphors* (M.W. Ho & S.W. Fox), pp. 63-84. Wiley, London,.

Prager, R., Beer, W., Voigt, W., Claus, H., Seltmann, G., Stephan, R., Bockemuhl, J. & Tschäpe, H. (1995). Genomic and biochemical relatedness between *vibrio-cholerae. Microbiol. virol. parasitol. inf. Dis.* 283: 14-28.

Provine, W.B. (1971). *The Origins of Theoretical Population Genetics*, Univ. of Chicago Press, Chicago.

Raven, P. (1994). Why it matters. *Our Planet* 6, 5-8.

Reddy, S.P., Rasmussen, W.G. & Baseman, J.B. (1995). Molecular-cloning and characterization of an adherence-related operon of myocplasma-genitalium. *J. Bacteriol.* 177: 5943-5951.

Redenbaugh, K., Hiatt, W., Martineau, B., Lindemann, J. & Emlay, D. (1994). aminoglycoside 3'-phosphotransferase-II (alph(3')II) - review of its safety and use in the production of genetically-engineered plants. *Food Biotechnology* 8: 137-165.

Regal, P.J. (1994). Scientific principles for ecologically based risk assessment of transgenic organisms. *Molecular Ecology* 3, 5-13.

Reidl, J. & Mekalanos, J.J. (1995). Characterization of *Vibrio-cholerae* bacteriophage-K139 and use of a novel mini-transposon to identify a phage-encoded virulence factor. *Molecular Microbiol.* 18: 685-701.

Rennie, J. (1993). DNA's new twists. *Scientific American* March: 88-96.

Reynolds, M.P., van Beem, J., van Ginkel, M. & Holsington, D. (1996). Breaking the yield barriers to wheat: A brief summary of the outcomes of an international consultation. CIMMYT.

Riede, I. (1996). 3 mutant-genes cooperatively induece brain-tumor formation in Drosophila malignant brain-tumor. *Cancer Genetics and Cytogenetics* 90, 135-141.

Rifkin, J. & Howard, T. (1980). *Entropy: A New World View*, The Viking Press, New York.

Riley, L.W., Remis, R.S., Helgerson, S.D., et al., (1983). Hemorrhagic colitis associated with a rare Escherichia coli Serotype. *New England J. of Medicine* 308, 681-85.

Ripp, S., Ogunseitan, O.A. & Miller, R.V. (1994). Transduction of a fresh-water microbial community by a new *Pseudomonas-aeruginosa* generalized transducing phage, *UTI. Mol. Ecol.* 3: 121-126.

Rissler, J. & Mellon, M. (1993). *Perils Amidst the Promise - Ecological Risks of Transgenic Crops in a Global Market*, Union of Concerned Scientists, USA.

Roberts, M.C. (1989). Gene transfer in the urogenital and respiratory tract. In *Gene Transfer in the Environment* (S. Levy & R.V. Miller, eds.), pp. 347-375, McGraw-Hill Book Co., New York.

Roberts, M.C. & Hillier, S.L. (1990). Genetic basis of tetracycline resistance in urogenital bacteria. *Antimicrobal Agents and Chemotherapy* 34, 261-264.

Rothenfluh, H.S. & Steele, E.J. (1993). Origin and maintenance of germ-line V genes. *Immunology and Cell Biology* **71**, 227-232.

Rowe, D. (1987). *Beyond Fear*, Fontana, London.

Roy-Burman, P. (1996). Endogenous env elements: partners in generation of pathogenic feline leukemia viruses. *Virus Genes* 11, 157-161.

Rubin, A.L., Arnstein, P. & Rubin, H. (1990). Physiological induction and reversal of focus formation and tumorigenicity in NIH-3T3 cells. *Proceedings of the national Academy of Sciences USA* 87, 482-486.

Rubin, H. (1992). Cancer development: the rise of epigenetics. *European J. of Cancer* 28, 1-2.

Salyers, A.A. & J.B. Shoemaker (1994). Broadhost range gene transfer: plasmids and conjugative transposons. *FEMS Microbiology Ecology* 15: 15-22.

Sandaa, R.A. & Enger, Ø. (1994). Transfer in marine sediments of the naturally occurring plasmid pRAS1 encoding multiple antibiotic resistance. *Applied and Environmental Microbiology* 60, 4243-4238.

Sandmeier, H. (1994). Acquisition and rearrangement of sequence motifs in the evolution of bacteriophage tail fibers. *Mol. Microbiol.* 12: 343-350.

Saunders, P.T. (1984). Development and evolution. In *Beyond neo-Darwinism: Introduction to the New Evolutionary Paradigm* (M.-W. Ho & P.T. Saunders, eds.), pp. 243-263, Academic Press, London.

Saunders, P.T. (1994). Evolution without natural selection: Further implications of the Daisyworld parable, *J. Theor. Biol.* 166, 365-373.

Saunders, P.T. (1997). *Evolving the Probable*, (to appear)

Saunders, P.T. & Ho, M.-W. (1995). Reliable segmentation by successive bifurction. *Bulletin of Mathematical Biology* 57, 539-556.

Saunders, J.R. & Saunders, V.A. (1993). Genotypic and phenotypic methods for the detection of specific released microorganisms, pp 27-59 in *Monitoring Genetically Manipulated Microorganisms in the Environment* (Ed. C. Edwards), John Wiley & Sons., New York.

Schäfer, A., Kalinowski, J. & Pühler, A. (1994). Increased fertility of *Corynebacterium glutamicum* recipients in intergeneric matings with *Escherichia coli* after stress exposure. *Applied and Environmental Microbiology* 60: 756-759.

Schluter, K., Futterer, J. & Potrykus, I. (1995). Horizontal gene-transfer from a transgenic potato line to a bacterial pathogen (*Erwinia-chrysanthem*) occurs, if at all, at an extremely low-frequency. *Bio/Techology* 13: 1094-1098.

Schneider, E.D. & Kay, J.J. (1994). Life as a manifestation of the second law of thermodynamics. *Mathl. Comp. Modeling* 19, 25-48.

Schnitzler, N., Podbielski, A., Baumgarten, G., Mignon, M. & Kaufhold, A. (1995). M-protein or M-like protein gene polymorphisms in human group-G *Streptococci*. *J. Clin. Microbiol.* 33: 356-363.

Schrag, S.J. & Perrot, V. (1996). Reducing antibiotic resistance. *Nature* 381: 120-121.

Schrödinger, E. (1944). *What is Life?*, Cambridge Univ. Press, Cambridge.

Schubbert, R., Lettmann, C. & Doerfler, W. (1994). Ingested foreign (phage M13) DNA survives transiently in the gastrointestinal tract and enters the bloodstream of mice. *Mol. Gen. Genet.* 242: 495-504.

Seitz, S., Rohde, K., Bender, E., Nothnagel, A., Kolble, K., Shlag, P.M. & Scherneck, S. (1997). Strong indication for a breast cancer susceptibility gene on chromosome 8p12-p22: Linkage analysis in German breast cancer families. *Oncogene* 14, 741-743.

Schihabuddin, L., Silverman, J.M., Buchsbaum, M.S., Seiver, L,J., Luu, C., Germans, M.K., Metzger, M., Mohs, R.C. Smith, C.J. Spiegelcohen, J. & Davis, K.L. (1996). Ventricular enlargement associated with linkage marker for schizophrenia-related disorders in one pedigree. *Molecular Psychiatry* 1, 215-222.

Shiva, V. (1993). *Monoculture of the Mind*,

Third World Network, Penang.

Shiva, V. (1994). Why we should say 'No' to GATT-TRIPS, *Third World Resurgence* 39, 3-5.

Shiva, V,, Jafri, A.H., Bedi, G. & Holla-Bhar, R. (1997). *The Enclosure and Recovery of the Commons*. Research Foundation for Science, Technology and Ecology, New Delhi.

Skogsmyr, I. (1994). Gene dispersal from transgenic potatoes to conspecifics: a field trial. *Theor. appl. Gene.* 88: 770-774.

Smirnov, V.V., Rudendo, A.V., Samgorodskaya, N.V., Sorokulova, I.B., Reznik, S.R. & Sergeichuk, T.M. (1994). Susceptibility to antimicrobials drugs of Baccili used as basis for some probiotics. *Antibiot-Khimiorec* 39, 23-28.

Sommer, S.S. (1995). Recent human germline muttion: inferences from patients with hemophilia B. *Trends in Genetics* 11, 141-147.

Soros, G. (1997). Capital Crimes, *The Guardian*, Saturday January 18, pp.1-3.

Sougakoff, N., Papadopoulou, B., Norman, P. and Courvalin, P. (1987). Nucleotide sequence and distribution of gene *tetO* encloding tetracycline resistance in *Campylobacter coli. FEMS Microbiological Letters* 44, 153-159.

Spallone, P. (1993). *Generation Games, Genetic Engineering and the Future for our Lives*, The Women's Press, London.

Speer, B.S., Shoemaker, N.B., & Salyers, A.A. (1992). Bacterial resistance to tetracyclne: mechanisms, transfer and clinical significance. *Reviews in Microbiology* 5, 387-399.

Spielman, A. (1994). Why entomological antimalaria research should not focus on transgenic mosquitoes. *Parasitology Today* 10, 374-376.

Spratt, B.G. (1988). Hybrid penicillin-binding proteins in penicillin-resistant strains of Neisseria gonorrhoeae. Nature 332, 173-176.

Spratt, B.G. (1994). Resistance to antibiotics mediated by target alterations. *Science* 264, 388-

Steele, E.J. (1979). *Somatic Selection and Adaptive Evolution*, Toronto.

Stephenson, J.R. & Warnes, A. (1996). Release of genetically-modified microorganisms into the environment. *J. Chem. Tech. Biotech.* 65, 5-16.

Stokes, H.W. & Hall, R.M. (1989). A novel family of potentially mobile DNA elements encoding site-specific gene-integration functions: integrons. *Molecualr Microbiology* 3, 1669-1683.

Stokes, H.W. & Hall, R.M. (1992). The integron in plasmid R46 includes two copies of the *oxa2* gene cassette. *Plasmid* 28, 225-234.

Strohman, R. (1994). Epigenesis: The missing beat in biotechnology? *Bio/Technology* 12,156-164.

Strohman, R. (1997). The coming Khunian revolution in biology. *Nature Biotechnology* 15, 194-200.

Summers, K.M. (1996). Relationship between genotype and phenotype in monogenic diseases: relevance to polygenic diseases. *Human Mutation* 7, 283-298.

Symonds, N. (1994). Directed mutation: a current perspective. *J. theor. Biol.* 169, 317-322.

Temin, H. M. (1980). Origin of retroviruses from cellular moveable genetic elements. *Cell* 21, 599-600.

Temin, H.M. (1985). Reverse transcription in the eukaryotic genome: retroviruses, Pararetroiruses, Retrotransposons, and Retrotranscripts. *Molecular Biology and Evolution* 2, 455-468.

Temin, H.M. & Engels, W. (1984). Movable genetic elements and evolution. In *Evolutionary Theory: Paths Into the Future* (J.W. Pollard, ed.), Wiley, London.

Thompson, W.C. & Ford, S. (1990). Is DNA fingerprinting ready for the courts? *New Scientist* 31 March, 38-43.

Thompson, M.W., McInnes, R.R. & Willard, H.F. (1991). *Genetics in Medicine*, 5th Ed., W.B. Saunders & Co., London.

Torres, O.G., Korman, R.Z., Zahler, S.A. & Dunny, G.M. (1991). The conjugative transposon Tn925: enhancement of conjugal transfer by tetracycline in *Enterococcus faecalis* and mobilization of chromosomal genes in *Bacillus subtilis* and *E. faecalis*. *Molecular and General Genetics* 225, 395-400.

Tschäpe, H. (1994). The spread of plasmids as a function of bacterial adaptability. FEMS *Microbiology Ecology* 15: 23-32.)

Trieu-Cuot, P., Gerbaud, G., Lambert, T., & Courvalin, P. (1985). *In vivo* transfer of genetic information between Gram-positive and Gram-negative bacteria. *EMBO J.* 4, 3583-3587.

Turing, A. M. (1952). The chemical basis of morphogenesis. *Philosophical Transactions of the Royal Society, London, B.* 237, 37-72.

Udo, E.E. & Grubb, W.B. (1990). Transfer of resistance determinants from a multiresistant *Staphylococcus* aureus isolate. *J. of Medical Microbiology* 35, 72-79.

Upton, M., Carter, P.E., Organe, G. & Pennington, T.H. (1996). Genetic heterogeneity of M-type-3 G group-A Streptococci causing severe infections in Tayside, Scotland. *J. Clin. Microbiol.* 34: 196-198.

van Heyningen, V. (1994). One gene-four syndromes *Nature* 367, 319-320.

Vasquez, G.R. (1997). Presentation in Workshop on "Protecting people's rights to productive resources", 22nd World Conference, SID, May, 1997.

Virgin, I. & Federick, R.J. eds. (1996). *Biosafety Capacity Building: Evaluation Criteria Development*, Stockholm Environment Institute, Stockholm.

Volk, W.A., Gerbhardt, B.M., Hammarskjold, M.L. & Kadner, R.J. (1995). *Essentials of Medical Microbiology*, 5th ed., Lippincott-Raven Publishers, Philadelphia.

Wahl, G.M., de Saint Vincent, B.R. & DeRose, M.L. (1984). Effect of chromosomal position on amplification of transfected genes in animal cells. *Nature* 307: 516-520.

Warren, A.M. & Crampton, J.M. (1994). Mariner: Its prospects as a DNA vector for the genetic manipulatin of medically important insects. *Parasitology Today* 10, 58-63.

Watkins, K. (1996). Free trade and farm fallacies. *The Ecologist* 26, 244-255.

Weatherall, D. (1993). *The New Genetics and Clinical Practice*, Oxford Univ. Press, Oxford.

Webster G. & Goodwin, B.C. (1982). On the origin of species: a structuralist approach. *J. of Social and Biological Structures* 5, 15-47.

Webster, G. & Goodwin, B.C. (1996). *Form and Transformation*, Cambridge Univ. Press, Cambridge.

Whatmore, A.M. Kapur, V., Musser, J.M. & Kehoe, M.A. (1995). Molecular population genetic-analysis of the enn subdivision of group-A-Streptococcal emm-like genes - horizontal gene-transfer and restricted variation among enn genes. *Mol. Microbiol.* 15: 1039-1048.

Whatmore, A.M. & Kehoe, M.A. (1994). Horizontal gene-transfer in the evolution of group-A Streptococcal emm-like genes - gene mosaics and variation in vir regulons. *Mol. Microbiol.* 11: 363-374.

Wilmut, I., Schnieke, A.E., McWhir, J., Kind, A.J. & Campbell, K.H.S. (1997). Viable offspring derived from fetal and adult mammalian cells. *Nature* 385, 810-813.

Wilson, E.O. (1975). *Sociobiology*, Belknap Press, Cambridge, Mass.

Winn, R.N., Vanbeneden, R.J. & Burkhart, J.G. (1995). Transfer, methylation and spontaneous mutation frequency of fX174am3cs70 sequences in medaka (*Oryzia-latipies*) and mummichog (*Fundulus-heteroclitus*) - implications for gene-transfer and environmental mutagenesis in aquatic species. *J. of Marine Environmental Research* 40: 247-265.

Wirz, J. (1997). DNA at the edge of contextual biology. in *The Future of DNA* (J. Wirz and E. Lammerts van Bueren eds.) pp. 94-103, Kluwer Academic Publishers, Dordrecht.

Wolfe, J. (1995). Cystic fibrosis. In *Processes of Heredity, S327 Living Processes Book 4* (S. Jones and K. Taylor, eds.), Open Univ., Milton Keynes.

Wolpert, L. (1996). In praise of science. In *Science Today: Problem or Crisis?* (R. Levinson and J. Thomas, eds,) pp. 9-21, Routledge, London.

Wright, H., Carver, A., & Cottom, D. (1991). High-level expression of active human alpha-1-antitrypsin in the milk of transgenic sheep. *Biotechnology* 9,

830-834.

Yin, X. & Stotzky, G. (1997). Gene transfer among bacteria in natural environment. *Applied Microbiology* (in press). (Preprint kindly supplied by the author, S.G.)

Young, R.M. (1985). *Darwin's Metaphor*, Cambridge Univ. Press, Cambridge

Glossary

Adaptive mutation, or **Directed mutation** The phenomenon whereby bacteria and yeast cells in stationary (non-growing) phase, have some way of producing (or selectively retaining) only the most appropriate mutations that enable them to make use of new substrates for growth.

Allele A particular variant of a gene.

Allergen A substance that causes the body to react hypersensitively to it.

Autosome A chromosome other than the sex-chromosome, or sex-deter-mining chromosome.

Bacteriophage Any virus that infect bacteria, also known as **phage**.

Biofilm A layer of extracellular matrix containing quiescent, non-proliferating micro-organisms.

Chromosome A structural unit of genetic material consisting of a long molecule of DNA complexed with special proteins in eukaryotes, but not in prokaryotes

Clone An identical copy of an individual or a gene, or the totality of all the identical copies made from an individual or a gene. In genetics, the clone is identical in genetic make-up to the original.

Conjugation The mating process in bacteria which require cell to cell contact being established, and in which genes are transferred between cells.

Directed mutation - See **Adaptive mutation**.

DNA DeoxyriboNucleic Acid, the genetic material consisting of a long chain of individual units called nucleotides, each consisting of a base joined to a sugar and a phosphate group.

DNAse An enzyme that breaks down **DNA**.

DNA methylation An endogenous process in the cell which adds a methyl group, -CH_3, to the base cytosine or adenosine, resulting in gene-silencing, or failure of the gene to become expressed.

DNA polymerase An enzyme that makes **DNA**.

Dominant allele An **allele** which is expressed when only one copy is present in an individual, i.e., in heterozygous condition.

Ecosystem The totality of all plant and animal species that constitute an interdependent, interrelated community.

Entropy A measure of the disordered, degraded energy that is unavailable for work.

Enzyme A **Protein** produced by living organisms that acts as a catalyst for a specific biochemical (metabolic) reaction.

Epistasis Interaction between genes.

Eukaryote The major class of living things, including all multicellular, higher organisms and some single-celled organisms, that have a **nucleus** in their cells, containing the chromosomes.

Gene A unit of heredity, usually a stretch of DNA with a well-defined function, such as one coding for a protein, or one that promotes transcription of other proteins.

Gene amplification The process whereby genes or a sequence of DNA in the genome is greatly increased in number of copies.

Gene cloning The technique of making many copies of a gene, isolating the gene and identifying it.

Gene expression In molecular genetics, this usually means the eventual appearance of the polypeptide encoded by the gene.

Gene silencing The process(es) whereby certain genes in the genome are prevented from being **expressed** by chemical modifications and other means.

Gene therapy Treating diseases by replacing the defective gene, either by incorporating a normal copy of the gene in the germ-cells (egg or sperm) or in the embryo (germline gene replacement therapy), or by supplying copies of the normal gene to be taken up and incorporated into cells of the adult (somatic cell gene replacement therapy).

Genetic code The code establishing the correspondence between the sequence of bases in nucleic acids (DNA and the complementary RNA) and the sequence of amino acids in proteins.

Genetic determinism Determinism is the doctrine that all acts, choices and events are the inevitable consequence of antecedent sufficient causes. Genetic determinism is the doctrine that the organism is the inevitable consequence of its genetic makeup, or the sum of its genes.

Genetic engineering The manipulating of genetic material in the laboratory. It includes isolating, copying, and multiplying genes, recombining genes or DNA from different species, and transferring genes from one species to another, bypassing the reproductive process.

Genetic marker Any segment of DNA that can be identified, or whose chromosomal location is known, so that it can be used as a reference point to map or locate other genes. Any gene which has an identifiable phenotype that can be used to track the presence or absence of other genes on the same piece of DNA transferred into a cell.

Genome The totality of the genetic material of a cell or organism.

Genotype The precise variant(s) of the gene(s) carried by an individual.

Heterozygote An individual who has two different alleles of a gene.

Homozygote An individual who has two identical allells of the gene.

Horizontal gene transfer Transfer of genes from one individual to another, of the same or different species, usually by means other than cross-breeding.

Interrupted genes Genes whose coding sequence is interrupted at intervals by long stretches of non-coding sequences. The coding regions came to be known as *exons* and the non-coding regions as *introns*. This structure is now found to be characteristic of most eukaryotic genes. The number and size of introns vary greatly, and they are often much longer than the coding sequences. After transcription, the intron regions are removed, or *spliced* out form the RNA transcript before it is translated into protein.

Messenger RNA The RNA intermediate in protein synthesis containing a transcribed copy of the gene sequence that specifies the amino acid sequence of the polypeptide it encodes.

Metabolism The sum total of the chemical processes that take place in living organisms, resulting in growth, development, and all other forms of energy transformation.

Metabolite One particular chemical intermediate generated in metabolism.

Mitochondria Membrane-bound cellular organelles in which organic substrates derived from food are oxidized to provide energy for all kinds of vital activities. They carry their own complement of DNA and are replicated independently so that when the cell divides, each daughter cell will receive half of the mitochondria.

Mobile Genetic Element, also called **Transposon** or **Transposable Genetic Element** A sequence of DNA that can move (transpose) from one place to another in the genome of a cell.

Multigene families are genes that exist in multiple copies in the genome, from several copies to many thousands or hundreds of thousands of copies.

Mutagen A substance or agent that causes genetic mutations, or chemical alteration of the genetic material, DNA.

Nucleus A structure in the eukaryote cell bounded by a membrane, that contains the genetic material, in the form of DNA organized into chromosomes.

Oncogenes Genes associated with cancer.

Pathogen Any agent that can cause disease.

Phage A bacterial virus

Phenotype The expressed characteristics, or an expressed character of an organism due to its genotype.

Plasmid A piece of parasitic genetic material found in a cell that can propagate itself using the cell's energetic resources.

Polypeptide or **Protein** A long chain of different amino-acids joined together by special chemical (peptide) bonds.

Polygenes The (hypothetical) many genes affecting a character, each having a small, additive effect on the character.

Prokaryote The class of living things, including all bacteria, that do not have a nucleus in their cell.

Promoter The region of a gene at which the RNA polymerase binds to start transcription. Most promoters are located upstream of the gene, except some eukaryotic genes which have promoters internal to the gene.

Proto-oncogenes Cellular genes which,

when mutated, or over-expressed become **oncogenes**.

Provirus A virus that has inserted its genome or a complementary copy of its genome into the host cell genome.

Recessive allele An allele which is not expressed unless two copies are present in the individual, i.e., in homozygous condition.

Recombination The formation of new combinations of alleles or new genes which occur when two homologous DNA or chromosomes break and exchange parts.

Reductionism The doctrine that complex systems can be completely understood in terms of its simplest parts. For example, an organism is to be completely understood in terms of its genes, a society in terms of its individuals, and so on.

Retrotransposon A mobile genetic elements that depends on a reverse transcription step to move and to duplicate.

Reverse Transcription The reverse of transcription - making a copy of complementary DNA (cDNA) from an RNA sequence - catalyzed by the enzyme, reverse transcriptase.

Ribosome An organelle in the cell required for protein synthesis.

Ribosomal RNA RNA molecules which make up the **ribosome**.

RNA RiboNucleic Acid, similar to DNA except for the sugar in the nucleotide unit, which is ribose, instead of deoxyribose, and the base Uracil instead of Thymine. RNA is the genetic material for RNA viruses,

RNA editing The process in which the base sequence of the RNA transcript is changed by addition of bases to the RNA molecule or by chemical transformation of one base to another. This subverts the genetic information carried in the genes.

RNA polymerase An enzyme that makes **RNA**.

Somaclonal variation Genetic variations of plant cells arising in cell culture, due to enhanced genetic instability.

Substrate A chemical substance that takes part in a chemical reaction catalyzed by an enzyme.

Thermodynamics The branch of physics dealing with the transformation of energy, especially of heat and other forms of energy.

Transcription The process of making a complementary sequence of the gene sequence in the genome, which is either used directly, as in case of **Ribosomal RNAs** (rRNAs) and **Transfer RNAs** (tRNAs), or is further processed into the messenger RNA, which is translated into protein. The process is catalyzed by the enzyme known as DNA-dependent RNA polymerase.

Transcription factors Proteins in eukaryotes that regulate the transcription of other genes by binding to regulatory sequences of the gene, interaction with one another and with the RNA polymerase.

Transduction In genetics, the transfer of genes by viruses from one organism to another.

Transfer RNA RNA molecules which transfer specific amino acids to the messenger RNA so that the polypeptide it encodes can be synthesized.

Transformation In genetics, the transfer of genes by one organism taking up DNA belonging to another organism of the same or different species.

Transgenic organism An organism created by genetic engineering, in which one or more foreign genes have been incorporated into its genome.

Translation The step in protein synthesis in which the messengerRNA directs the synthesis of a polypeptide of a particular amino-acid sequence by 'decoding' the genetic code.

Transposon - See **Mobile genetic element**.

Virus A parasitic genetic element enclosed in a protein coat that can replicate in cells, and form infectious particles, or remain dormant in the cells. Its genetic material can become integrated into the cell's genome to form provirus.

Virulence Ability (of pathogens) to infect organisms and cause disease

Dr. Mae-Wan Ho

Mae-Wan Ho is that rare combination - a holistic biologist. A well-known and respected British scientist, she is Reader in biology at the Open University, UK, where she teaches and researches genetics, evolution, and the physics of organisms and sustainable systems. Dr. Ho gained her PhD in biochemistry at Hong Kong University. She became a post doctoral Research Fellow at the University of California in San Diego and was awarded a Fellowship of the National Genetics Foundation, USA. She went on to become a Senior Research Fellow at the University of London before joining the Open University in 1976. In addition, she is a popular public lecturer and frequent contributor to radio and TV in the UK and elsewhere.

Mae-Wan Ho is also a prolific writer with more than 150 publications across many disciplines. Since 1994, she has been scientific adviser to the Third World Network and other public interest organisations on genetic engineering biotechnology and biosafety. She has written key papers for the public and for policy-makers and debated with spokespersons in the biotech industry. This recent experience and her wide-ranging expertise in many fields of scientific research make her uniquely qualified to write this book.

Do Ho's previous books include *Beyond neo-Darwinism: Introduction to the New Evolutionary Paradigm* (co-edited with P. T. Saunders), 1984; *Evolutionary Processes and Metaphors* (co-edited with S. W. Fox), 1988; *Bioenergetics*, 1995; *The Rainbow and the Worm: The Physics of Organisms*, 1993 & 1998; a forthcoming book *Genetic Engineering and the Gene Ecology of Infectious Diseases*, 1998.

Born in November, 1941, Mae-Wan Ho is married and lives in London.

Index